Quality Systems and Controls for Pharmaceuticals

Quality Systems and Controls for Pharmaceuticals

Dipak K. Sarker

School of Pharmacy and Biomolecular Sciences
University of Brighton, UK

John Wiley & Sons, Ltd

Copyright © 2008   John Wiley & Sons Ltd, The Atrium, Southern Gate, Chichester,
West Sussex PO19 8SQ, England

Telephone (+44) 1243 779777

Email (for orders and customer service enquiries): cs-books@wiley.co.uk
Visit our Home Page on www.wileyeurope.com or www.wiley.com

Other Wiley Editorial Offices

John Wiley & Sons Inc., 111 River Street, Hoboken, NJ 07030, USA

Jossey-Bass, 989 Market Street, San Francisco, CA 94103-1741, USA

Wiley-VCH Verlag GmbH, Boschstr. 12, D-69469 Weinheim, Germany

John Wiley & Sons Australia Ltd, 42 McDougall Street, Milton, Queensland 4064, Australia

John Wiley & Sons (Asia) Pte Ltd, 2 Clementi Loop #02-01, Jin Xing Distripark, Singapore 129809

John Wiley & Sons Canada Ltd, 6045 Freemont Blvd, Mississauga, Ontario, L5R 4J3, Canada

Wiley also publishes its books in a variety of electronic formats. Some content that appears
in print may not be available in electronic books.

Library of Congress Cataloging-in-Publication Data

Sarker, Dipak K.
   Quality systems and control for pharmaceuticals / Dipak K. Sarker.
     p. ; cm.
   Includes bibliographical references and index.
   ISBN 978-0-470-05692-9 (cloth : alk. paper) – ISBN 978-0-470-05693-6 (pbk. : alk. paper)
   1. Pharmaceutical industry–Quality control. I. Title.
   [DNLM: 1. Pharmaceutical Preparations–standards–Handbooks.
   2. Drug Industry–standards–Handbooks. 3. Quality Control–Handbooks.
   4. Technology, Pharmaceutical–standards–Handbooks. QV 735 S245q 2008]
   RS189.S27 2008
   338.4′76151 – dc22

                            2008016829

British Library Cataloguing in Publication Data

A catalogue record for this book is available from the British Library

ISBN 978-0-470-05692-9 (HB)   ISBN 978-0-470-05693-6 (PB)

Typeset in 10.5/13.25 Times by Laserwords Private Limited, Chennai, India

This book is printed on acid-free paper

Dedicated to my son
Hugh Callum Sarker

Contents

Preface

Objectives

This book hopes to answer some of the questions asked by undergraduates and industrialists that are coming to the area of industrial pharmaceutics, pharmaceutical science and industrial pharmaceutical practice. The subject matter can at times be terse, full of jargon and rather dry, but also intrinsically diverse, and an appreciation requires a solid grounding in subjects covered at lower level during a diploma or degree programme.

Style and format

The text hopes to balance size with practicality and cost and therefore accessibility to non-experts in the subject and tries not to do everything (a current problem with some texts), but as a must, covers the ground that is part of quality systems and pharmaceutical manufacture. Without a doubt each of the sections could be expanded greatly but that depth falls outside this text as a 'pocket-guide'. The principle aims of this text are to target undergraduates (foundation degrees, BSc pharmaceutical sciences), postgraduates (MPharm, MSc (specialisms), PgDip, etc.), industrialists, practicing clinicians, researchers and scientists making the transition to industry and to also serve in support of in-house short-course training events.

Some topics such as 'documentation' and 'auditing' are not given sections but are incorporated into other areas; these particular topics are covered in the quality assurance (QA) section, which comes under validation and manufacturing. Likewise 'raw materials' and 'microbiology' appear under GMP/hygiene and marginally under analytical testing. Process analytical technology (PAT) is covered under cGMP (current good manufacturing practice) and good laboratory practice (GLP). 'Biotechnology' appears under the title of biopharmaceuticals. I have been wary of alluding to a full discourse on biotechnology (a) because the science is changing fast now and (b) because this can be dealt with in more detail elsewhere. Therefore I could not do justice to this subject in a limited section in a book designed to provide an essential overview.

The emphasis of the text is on science and regulation but is weighted toward quality systems (TQMS and cGMP). Unfortunately there is no way of escaping the 'currency of regulation information provided', so since the time of going-to-press some of the legal stipulations and requirements may have changed. Regulation moves fast, but in some sense not that fast, and the essentials remain in practice, although terminology and definition may change! My aim has been to produce a better 'quality system manual' and course text for which there is no 'realistic alternative' in the sense that I, like many other lecturers, use a wide range of disparate texts to deliver this lecture material. General and more specialised pharmaceuticals are discussed in depth roughly equal to their relative, technical make-up and societal/medical importance. The book makes frequent reference to current data and websites (that may change with time) of many official bodies (FDA/MHRA/EMEA), which industry professionals and practicing pharmacists use to access the compliance information used routinely in pharmaceutical mass manufacture.

The book also includes best practice examples, for which I draw on my industrial experience and discussions with practicing industrial pharmacists that hopefully contemporise this issue. The book is intended as a study guide and so also includes examples of essays and seminar topics; one good example being good clinical practice (cGCP). This comes in under the section on 'new product development' that talks about the development and product life cycle (pharmaco-economics). Again this is weighted to phase IV of the development process and pre-launch status/routine manufacture. I think too much reference to clinical trials is not too helpful or suitable if the text is meant to focus on good manufacturing practice. I also hope to whip up support among students and clearly demonstrate the relevance of quality systems in the subject areas directly involved in industrial pharmacy, since this approach extends to all other areas of pharmacy practice. Community pharmacy tends to dominate the pharmacy and pharmaceutical sciences spectrum, without a stress placed on the underpinning QA support to hospital and community pharmacy; this is certainly the case at my university and undoubtedly holds true elsewhere. I hope the text will prove to be useful and serve as a point of reference.

Acknowledgements

I wish to acknowledge the helpful comments and suggestions of the Wiley team and the reviewers whom improved the manuscript at various stages. I also wish to acknowledge the helpful comments, advice and informal discussion of friends and colleagues both academic and industrial that helped me formulate the text. I would particularly like to thank my wife Ralitza and my mother Brenda for their encouragement.

Dipak Sarker

List of figures

List of tables

Glossary of terms and acronyms

α – probability of a type-I or producers' error

AOQL – average outgoing quality level

API – active pharmaceutical ingredient, active, drug

AQL – acceptable quality level

β – probability of a type-II or consumers' error

cGMP – current good manufacturing practice

CIP – clean in place

CoA – certificate of analysis

COV – coefficient of variation

CPMP – Committee for Proprietary Medicinal Products

CRM – certified reference material

CuSum – cumulative sum

DDS – drug delivery system

DMAIC – define, measure, assess, improve, control (part of the 6-sigma method)

EPT – end product testing

FDA – Food and Drug Administration (USA)

GCP – good clinical practice

GLP – good laboratory practice

GXP – good overall practice

HEPA – high energy particulate air (clean air)

HVAC – heating, ventilation and air-conditioning (system)

ICH – International Conference on Harmonisation (of medicines)

IND – investigational new drug

IPC – in-process control

IPR – intellectual property rights

IQ – installation qualification

ISO – International Standardization Organization

IUPAC – International Union of Pure and Applied Chemistry

LTPD – lot tolerance percent defective level

MHRA (MCA) Medicines and Health Regulation Agency (formerly Medicines Control Agency)

MII – maintain, improve or innovate

NCE – new chemical entity

NDA – new drug application

NFD – normal frequency distribution, normal distribution

NPD – new product development

OCC – operating characteristic curve

OQ – operational qualification

PCQ – purity, consistency and quality

PDCA – plan, do, check, act (Shewhart (Deming) cycle)

PIC – product innovation charter

PQ – performance qualification

QFD – quality function deployment

QP – qualified person

SDU – safety, dosage and usefulness/use (of a new drug)

SEM – standard error of the mean (variation, see precision)

SOP – standard operating procedure

SRM – standard reference material

TPP – target product profile

TQC – total quality control

TQM – total quality management

UAL – upper action limit

UQL – unacceptable quality level

USP – United States Pharmacopeia

UWL – upper warning limit

VMP – validation master plan (method)

VMR – validation master report (conclusion)

VP – validation plan

Glossary of mathematical and statistical symbols

A, k – constants

d, D – number of defectives

θ – defined angle (CuSum charts)

Me – median average

Mo – modal average

N – lot size

n – sample size

P – probability

R – range

R^2 – correlation coefficient

σ_{hat} – estimate of true population mean

σ_s – standard deviation of sample (see 6-sigma systems)

s – sample standard deviation

s^2 – sample variance

μ – population mean (average)

μ_1 – bias, a cause of error in estimations and across validations (systematic error)

x_{bar} – numerical average of samples (control chart centreline value)

x_{db} – average of individual sub-group averages

Other symbols are explained at their point of use.

Glossary of mathematical and statistical symbols

SECTION A

Most Suitable Environment

1

Introduction

This book and the nebulous area of science it belongs to are based on the evaluation and concept of assuring quality and good practice. This knowledge is routinely employed in the safe and hygienic manufacture of pharmaceuticals (medicines), cosmeceuticals (cosmetic-pharmaceuticals), and nutraceuticals, (nutritional-pharmaceuticals). However, the subject area is diverse and might also routinely apply to those aspects of pharmaceutical manufacture that are intimately associated with production, such as process control testing. Equally the subject matter might be relevant to disparate industries and environments such as the hospital histo-pathology lab, clinical biochemistry lab, cosmetics and semi-conductor industries, or paints, pigment and dye product industries, to name but a few. The basic elements of routine production of a non-exhaustive list of pharmaceutical products are shown in Figure 1.1. Three elements are key: the raw material (RM), process, and human intervention. The way in which these components interact or rather *are made to interact* for a range of products such as solid dosage (tablets) and dispersions, for example vaccines, defines their compliance, safety and ultimate suitability.

The full picture of drug formulation is complex, by necessity, and dictates the efficacy of the drug product in addition to its universality of use, application, regulatory status and need for careful administration. This is not the complete picture of drug product because successful medicines can only be made by using a complementary mix of paradigm models of 'quality' practice and scientific advancements [Sharp (2002)]. These quality models are essential and have evolved, having been borne out of a key craft and essential skills in the distant past, and are now defined by technological progress and know-how (Figure 1.2). The vast array of medicines and their appropriate engineering for purpose also comes from both empiric discovery and a rational design of medicines.

Quality Systems and Controls for Pharmaceuticals D K Sarker
© 2008 John Wiley & Sons, Ltd

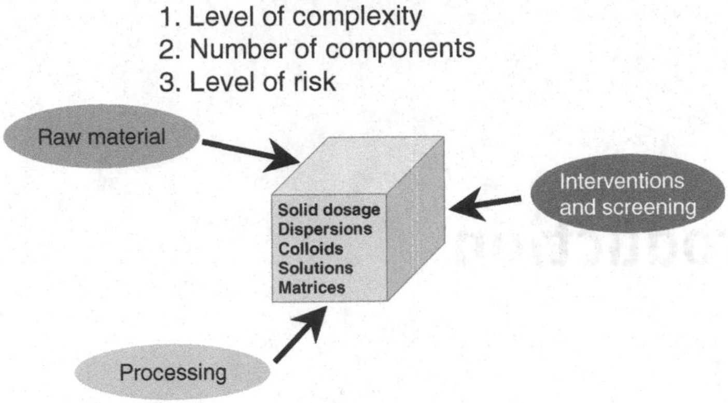

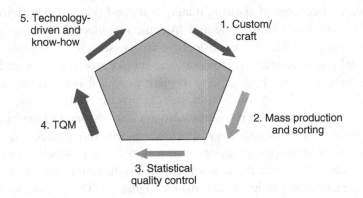

1. Level of complexity
2. Number of components
3. Level of risk

Raw material

Solid dosage
Dispersions
Colloids
Solutions
Matrices

Interventions
and screening

Processing

Figure 1.1   Pharmaceutical products overview

5. Technology-
driven and
know-how

1. Custom/
craft

4. TQM

2. Mass production
and sorting

3. Statistical
quality control

Figure 1.2   Paradigms of industrial quality practice

1.1 The process of finding new lead medicines

Recent drug development history begins with opium discovery in 1806. It was a most significant, ground-breaking discovery that gave rise to morphine and its subsequent associated compounds used ubiquitously since Victorian times for pain relief; other examples from the same era include quinine from Cinchona tree bark, revealed at the turn of the nineteenth century [Schacter (2006); Rang (2006)]. Serendipitous discovery features to a significant extent alongside empirical and purposeful experimental design in discovery; for example analgesics such as aspirin from willow; theobroma oil from cocoa butter, used for body temperature melting waxes; and cardiotonics, for example digitalis. The classic, often quoted example of chance discovery is that of the mycotoxin penicillin (*P. notatum*) in 1928 by Fleming, used for groundbreaking therapy from the 1940s, and a whole genre of new engineered medicines, drug discovery culture and biotechnology.

A number of serendipitous discoveries have included disclosures identified from *favourable* side-effects:

- Mogodon (nitrazepam) – routine use: hypnotic and sedative; second use: anti-emetic

- Tamoxifen – routine use: anti-rheumatic; second use: anti-oestrogenic (breast neoplasm)

- Aspirin – routine use: analgesic; second use: anti-coagulant

- Pyridoxine – routine use: enzyme cofactor; second use: serotonin inhibitor (depression disorders)

- Alginate – routine use: red seaweed/polysaccharide, excipient in tabletting; second use: acid reflux retardant

- Minoxidil – routine use: cardio-therapeutic; second use: hair growth (re-growth).

A number of fortunate discoveries have also included scientific findings by *chance*:

- Antibiotics such as myxins (*Sorangium* spp.); cephalosporins (*Cephalosporium* spp.)

- Toxins e.g. melittin (polypeptide) – in honey bee venom, used as anti-rheumatic; curare – Chondrodendron bark, used as muscle relaxant

- Codeine, from the giant poppy (diamorphine/morphine) – used for pain relief and as a sedative

- Quinine, from cinchona bark – used as an anti-malarial

- Salicylates, from willow bark – used as an analgesic

- Digitalis, from foxglove plant – used as a cardiotonic

- Cannabinols – from hemp bush – used as an analgesic or hypnotic drug

- Atropine – from deadly nightshade plant (belladonna) – used as an anti-cholinergic.

Clinical and pre-clinical testing and file submission for a lead compound to candidate takes about 10 years. Revolutionary cytotoxics such as Taxol® (paclitaxel) from yew trees, cis- and trans-platinates/oxyplatinates (colon cancer), are examples of inorganic therapeutics valued for treatment of cancers and are other novel and innovative classes. Drugs now used for alternative therapy (to the original filing) as a result of side-effects include e.g. Minoxidil, now used as a hair restorative, and thalidomide, which keeps on finding new applications, other than the original use as a sedative. One of the most controversial and criticised of contemporary cosmetic drugs in use is 'botox', botulin and its derivatives, a powerful neurotoxin, which now relies on routine 'biotech' fabrication. Newer aspects of medicinal delivery include stealth preparations and prodrug moieties to avoid first pass degradation and permit active delivery; typical examples include PEGylated liposomes and peptides, such as insulin. This is also a strategy widely used to avoid the blood brain barrier (BBB) and target tumours [Tian *et al.* (2005); Wang *et al.* (2005); Thomas and Campbell (2004)]. This paints rather a glorious picture of unflawed and successful progression of lead compound to drug and drug product. In 2005 there were about 75 new potential biopharmaceutical candidate drugs. Many will fail due to the rigorous burden of testing placed on them. Even successful drugs are not without their drawbacks, for example three products or combinations used in cancer treatment by inhibition of tyrosine kinase. These very successful products include Trastuzumab-Herceptin™ (Genetech) used with the cytotoxin adriamycin for breast cancer metastasis, Gleevec™ (Novartis) for chronic myelogenous leukaemia, and Sutinib Malate-Sutent™ (Pfizer) for metastatic renal and gastric carcinoma; however each silver lining has a potential black cloud as they are reported [Mann (2006)] to have significant administration side effects including possibilities for heart damage, heart failure and heart ventricular dysfunction, respectively.

1.2 A drug discovery framework

Pharmaceutical innovation is really big business, for example in the USA in 2005 it involved half a million people and the spending of $30 billion on research and development, and this related to more than $200 billion sales in North America. It has been estimated that in the US the top ten strategic targets for illnesses in 2004 were cancer (32 per cent), diabetes (9 per cent), arthritis (7 per cent), infections (6 per cent), HIV (4 per cent) and cardiovascular ailments (4 per cent), respectively. The business is healthy and investment is growing based on global sales, growing from $590 billion in 2003 to an astonishing $900 billion in 2008 (prediction based on current growth patterns). The biopharmaceutical market alone was $50 million in 2005 accounting for some 25 per cent of all drug candidates. Of these sales it is estimated that over-the-counter (OTC) will account for 11 per cent, generic medicines for 7 per cent, biopharmaceuticals also 7 per cent and ethical medicines have the greatest share at 75 per cent. In 2003 the top ten pharmaceutical companies accounted for 46 per cent of global sales and of the sale 50 per cent were in North America. In recent years the costs have forced investment down in absolute terms but in relative terms there is something like a two-fold increase in investment each five years [Tambuyzer (2002); Carpe Diem Publishers (2004); Mudhar (2006)]. Current drug discovery (Figure 1.3) also makes use of

Step	Activities	Significance
(a) Start point – disclosure	Target compound identified	
(b)	Strategic compound identified by organisation	
(c)	Pre-clinical screening tests e.g. in animal	A 'lead' molecule of pharmaceutical value leads to (d)
(d) Permission of human tests for IND	List value, threats, deficiencies, urgency (short referential tests)	
(e) True clinical trials (SDU)	Phase I, phase II, phase III (long, extensive and detailed)	Success here: organisations are looking to phase IV and V testing
(f)	Make filing of NDA	
(g)	Authority review process e.g. direction from EMEA, FDA*	
(h)	Approval of NCE (long, arduous)	
(i)	Phase IV and industrial commercialisation steps#	
(j) Finish point – product	Launch, commercial review	This is not the end of the line for the product
(k) Post-approval activities	Phase V	Periodic review based on clinical (e.g. GP) data

Key:
NCE – new chemical entity; new lead compound
IND – investigational new drug
NDA – new drug application
SDU – establishment of safety-dosage-use comes from clinical trials.
– many steps here and the economic and practical considerations mean some candidate molecules are lost.
* – further down the development cycle list means a greater degree of investment by the organisation and also the risk of failure or at least the consequences of failure are multiplied. Different bodies within the organisation may review the drug based on its chemical nature e.g. biologic, human or veterinary pharmaceutical class.

Figure 1.3 Process flow showing the fundamentals of making new medicines: there are 11 key steps, although many more sub-steps

chance discovery and not just rational experimental design and, consequently, many scientists in the field are still using the notion of miracle cures from natural sources as a basis of new drug products. This is and will continue to be part of the culture of drug discovery. However, increasing use of rational experimentation, bioinformatic profiling and high throughput screening (HTS) for analysis or pharmacologic/pharmacokinetic testing and early stage candidate screening is finding an ever more prominent position (Figure 1.4). This stands to reason, after all why put all your 'eggs in one basket' and hope for a miracle discovery.

Clinical trials and the integrity of clinicians and analytical biochemistry, pharmacology and chemometric aspects of clinical testing are crucial. All tests require an environment of good clinical practice (GCP) and this is explained more

Phase I	Where small-scale studies* reveal working dose and safety (immunity compromising drugs not tested in 'healthy' patients). Primary goal is evidence gathering for drug activity.
Phase II	At this point a more significant large group of models are evaluated in different types of trials (with statistical relevance e.g. double-blinded, random, placebo, etc.) to prove the molecule is efficacious and safe. Cost can be high as testing is performed in diverse testing centres (c. $50 million).
Phase III	Measurements undertaken in a heavily scrutinised and controlled environment over diverse sites (to prevent bias) with an extensive number of volunteers; the candidate molecule is tested in terms of SDU and toxicology and in replicate evaluations. Their overarching aim is to demonstrate utility in particular groups and provide a basis for marketing and supplementary file validation information (c. $120 million, [Schacter (2006); Rang (2006)])

* Small scale does not mean limited but rather of a survey type, rather than proof-of-efficacy.

Figure 1.4   What is the true picture of processes involved in clinical trialing?

in terms of generic components in the section on current good manufacturing practice (cGMP), in Section 5.

Less than a quarter of drugs at phase I entry will progress to reach NDA status. The development system is fraught with problems of burdensome cost (clinical trial cost up to $10 000 per patient) and the extent of randomised uncontrolled study. This means the cost of producing a new medicine can run to about $1 billion and the process can appear to be painfully slow, taking in many cases more than a decade to gather regulatory support and subsequent approval for use of a new drug. The regulators find themselves in an unenviable position of needing to prove the new drug will be 'sufficiently' risk free. Yet, based on a number of ongoing and past difficulties, find an increased public scrutiny and pressure to release new and exciting drugs quickly and not stifle creative innovative therapy.

Phase III trial data forms the main argument in favour of progression of the drug to pre-launch phase IV viability studies. Suitable information must be generated prior to clinical studies on a near optimum form of presentation and suitability of production of the intended candidate molecule for best results. This presupposes the organisation operates within an ongoing cGMP and cGLP quality system, which operates from the point of discovery. This production development goes on in the background and can take approximately 10 years with each of the other overlapping steps in the process and each of the clinical trial portions and filing or approval taking between one year and two years. It is not uncommon for 70 per cent of the time for intellectual property rights for a new substance to be consumed by development and approval. In most cases patent lifetime is of the order of twenty years but this varies on a case-by-case basis and with the country of filing.

The backbone of drug development is based around clinical acceptability of the investigational drug application (IND) compound. The trials move progressively to more rigorous and in-depth particularities of the candidate molecule when in human models. This is often also considered to be the most contentious part of the development cycle as there needs to be 'real proof of principle.' This consequently leads to some notion of an equilibrium of competing forces between input and provision of 'full information' or fact by developer and the regulator. Both parties have a vested interest in their successful participation and more often than not these interests are mutually compatible. After successful review of data by expert regulators the candidate molecule moves from IND class to one of a new drug application (NDA) class.

Development of new medicines should consider:

- Development of drugs and their societal value

- Profiling of drugs with respect to the potential market/customer

- Profiling of drug products with respect to the potential market/customer

- Specific design for a targeted application.

Alternatively new, improved or better drugs may be developed by planning and rational design. In this case key considerations should be:

- Use of disease prevention (retardation) strategies and models

- Use of chemical libraries or alternatively by new drug synthesis (empiric discovery)

- Use of pharmacological site targeting based on e.g. drug chemical structure

- Use of drug selection or screening (*in vivo/in vitro*) and bioassay (empiric discovery)

- Evaluation for favourable side-effects (not a main form of discovery) and chance by taking drugs from natural sources (e.g. plant extracts)

- Looking for more efficient yields and synthesis cost. High cost is more likely to reduce or limit the potential use of the drug

- Use of drug and product profiling (requirements of a drug such as target product profile)

- Use of the clinical trial and appropriate toxicological surveillance

- Use of placebos in product validation and random/blind/sequential trials aligned against the validity of clinical data [Tambuyzer (2002); Benoliel (1999)]

- High throughput screening and evaluation based on drug absorption, metabolism, distribution and elimination

- Consideration of bulk manufacture and process stability.

Some specific design considerations, which reduce the likelihood of new drug molecule acceptance, are:

- Risk of chemical or bio-mutagenesis
- Drug–excipient interactions
- Drugs involving poisoning of receptors e.g. stereospecific fouling and inhibition
- Drug delivered by a non-receptor pathway
- The time course of drug action
- Non-conventional modes of delivery that might require widespread novel clinical trials data to support their use
- The need for effective administration and its drug delivery ratio (delivered dose/absorbed dose)
- The risk of DDS toxicity and allergy.

2
Technology transfer and the climate of change

At any moment in time the current status of an acceptable quality of product is constantly redefining its position and workers in the area are constantly looking to use the latest developments from industry, academia and clinical practice to create the most suitable manufacturing environment. Pharmaceutical products vary immensely in terms of their complexity, including the number of components, depth of processing required and screening or intervention and testing or scrutiny required to ensure safety. Accepting that faults are made and that there is a scope for improvement is now an *ethos* widely absorbed into industrial, clinical and research practices. The key point is to act on diagnosis of mistakes and institute an improvement. Historically this has taken place as part of the academic peer review process.

The internal and external customers referred to in Table 2.1 simply relate to sources where new state-of-the-art findings can find a first point of application. There are two clear points where innovation can be drawn out of the development process, at stage 3 and stage 5. Earlier exploitation may provide opportunities for competitive advantage and real opportunities for technology – rather than market-driven change.

2.1 Innovation and research

All research-based disciplines, such as pharmacy and 'pharmaceutical sciences' are subject to an overarching strategy called a Quality Management System (QMS). Most of the details and issues considered in this text will at some point

Quality Systems and Controls for Pharmaceuticals   D K Sarker
© 2008 John Wiley & Sons, Ltd

Table 2.1   Research and development: total technology transfer

Stage	Process	Making use of 'blue-skies' or grounded research
1 – Initialisation	Academic *study*/ research theme	
2	Technology transfer • Internal customer • External customer	When, where, how and why this takes place depends on appropriate staffing and the qualities of the technical team
3	Peer review process (internal and external audit)	After sufficient review, qualified data/findings decanted to industry – innovation for further trials
4	Discussion and validation	
5	'Quality' findings	Fully qualified technology
6 – Possible exit point of academic loop	Routine application	Improvement without innovation

refer to a QMS organised at some level of sophistication. In the simplest analogy the components of a QMS are:

1. Research and Development (R&D)

2. Quality Assurance (QA).

Of course the picture is nowhere near as straightforward as this because both R&D and QA must consist of a number of functions and subgroups in order to work effectively. However, it is the correct amalgamation of the two that gives rise to product or outcomes of appropriate quality [Hoyle (2006)].

There can be no notion of an appropriate standard or benchmarking without a concept of total quality. The three golden rules of good practice are:

1. *Honesty* throughout all processes; strict ethical rules must apply including those against coercion, bias and profit-making.

2. *Rigour*; no process shall be deemed investigated fully with due care, attention and replication.

3. *Sound scientific basis*, meaning the underpinning logic or reason, strategy or method must be based on firm evidence and findings.

Upgrades in the working environment as a general rule result from increases in or more appropriate knowledge, better application of this know-how, chance

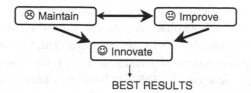

Figure 2.1   The 'maintain, improve or innovate' cycle (MII)

assignations and serendipity, and proper extensive and profound investigation. Any manufacturer striving for better assurance of quality (QA) must consider three possible courses of action. These are innovation, improvement or maintenance as shown in Figure 2.1, with innovation being considered the best option.

The three components fit into a cycle called the M2I or MII cycle because of their make-up [Hoyle (2006)]. Here, to maintain merely keeps the state of affairs as they are, improvement may feed back directly into QA and logically can be small or comprehensive. However, innovation provides the manufacturer or group with real competitive advantage [Anik (2002)] thus facilitating stronger perception or in real terms increased business possibly as one hopes by producing a superior product (Figure 2.2).

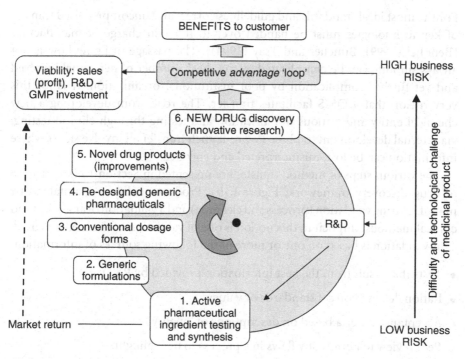

Figure 2.2   Competitive advantage in drug production and innovation. Adapted from [Amir-Aslani and Negassi (2006)]

Strategies for development can be based on using appropriate scientific models and modelling. Areas such as pharmacology, medicinal chemistry, quantitative structure-activity relationship (QSAR) prediction and high throughput screening (HTS) benefit from these areas of investigation. Significant social benefits from innovative new chemical entities (NCEs) have also been obtained via use of libraries and record re-tracing, taking radically new synthetic approaches and 'platform science'. Serendipity has been responsible for a host of groundbreaking medicines. In the 1800s morphine was identified from the opium poppy; it is used for pain relief to this day. Then quinine from *Cinchona* bark, since used as an anti-malarial; in 1909 arsenic was discovered as an anti-syphilitic by Paul Ehrlich, and in 1928 Alexander Fleming extracted antibiotic from *Penicillium notatum* [Schacter (2006)]. The latest findings have included Minoxidil, a cardiotonic now used for hair restoration; the nitric oxide inhibitor used for heart disease, now more famous for assisting male erectile dysfunction, as Viagra™; and Tamoxifen the anti-rheumatic now more well-known for breast cancer therapy. The list is seemingly endless and illustrates a point that the human body is complex and new drug discovered side-effects (secondary effects) *can* have potential novel applications of significant social value.

2.2 Method transfer

For the most ideal standards and guidelines correct and uncompromised transfer of key technologies must be handed over to those in charge of manufacture [Benoliel (1999); Buncher and Tsay (1994)]. The passage of know-how is not always smooth, can be complicated by the sheer number of personnel involved and yet further complication by poor institutional organisation. It is for this very reason that a QMS facilitates transfer. The road from developing a new chemical entity and notions of fabrication and testing through clinical testing, via internal development to phase IV pre-launch transfer all involve stages where information can be lost, misinterpreted and encoded.

The various steps of method transfer are presented in Section 10 (see cartoon of drug discovery framework, Figure 8.1b). Product quality is maintained by manufacturing validation (process and cleaning) and a validation of analytical and clinical methods. Implicit in this notion is one of validation of the methods used. This validation is based on one or more of the following sources of information:

- Tests that result from the best information provided by R&D

- Rationale based on set standards or values

- Acceptance criteria based on descriptions of 'quality'

- Peer review to permit any flaws in a process to be highlighted

- Feedback loop or iterative cycle of acting on findings.

3

Quality systems structure and a maximum quality environment

Talking of quality systems sounds a bit like trade jargon but in fact it is what it says it is: a system put in place to orchestrate quality through a process culminating in a product of the utmost quality. Here we need to change tack and first define what we mean by quality before going further. A number of definitions of *quality* exist:

1. '. . .degree of excellence possessed by an item. . .'

2. Meeting requirements of specific customer needs

3. Reliability contract with potential customer.

Definition 1 is straight from the Oxford English Dictionary but can be expanded to consider statistical framing using *3 or 6 sigma*. Definition 2 is considered to be the most valuable. At this point it is perhaps useful to expand definitions to core elements of TQMS.

- Quality assurance (QA) – a planned system of activities designed to ensure effective quality control. It always consists of the following central considerations:
  - Sampling, and good sampling based on solid statistics is crucial; two-thirds of all errors made during an investigation usually relate to unrepresentative sampling
  - Adequate testing, crucial as for sampling

- Responsibility for errors and decisions, normally a qualified person
- Checkpoints must be established by way of evaluation of potential risk.

- Quality control (QC) – a planned coherent system of activities designed to provide quality product.

A general model of controlling quality involves first applying a standard, then second implementing and checking the value or magnitude of the set standard, followed by checking the product for conformance, then instituting an appropriate remedy or action and feeding this back into the implementation and checking stage as an iterative improving cycle [Hoyle (2006); Kolarik (1995); Sharp (2002)]. Actual control of quality necessitates a more comprehensive involvement of all groups within the controlling departmental sections; this might typically involve in-process control, post-process control and finished goods control including stability testing. For this reason QC is often reported as being the more appropriate total quality control (TQC) concept. Inherent in this notion of total control are the following:

- Quality has to be *'built into'* the product (see the quality guru's work; Section 3.1) it cannot be built into or designed into a poor individual aspect of a process with an appropriate outcome.

- No other approach is acceptable except getting the product *'right-first-time'*.

- Faults and non-conformance are minimised by using the correct *environment*.

- The best outcomes derive from a sense of contribution across all sections and *team effort*.

- Best practice comes from a quality department that is *integrated with R&D*.

Finally, it is not possible to mention a quality filled corporate pharmaceutical culture without mentioning current good manufacturing practice (cGMP) and validation [Underwood (1995); Bourget *et al.* (2001); FDA (2001); FDA (2003); Moritz (2005); Mollah (2004)]. cGMP occurs when the process of manufacture is clearly defined. That is to say, that testing and processing methodology are assessed and optimised; this usually occurs through lengthy validation. Validation is an established regimen of activities based around process capability showing that a SYSTEM of PRACTICE does EXACTLY what it is supposed to do. We will define quality and process capability in mathematical terms later in the text (see Section 6.1) under 3 and 6-sigma, based on normal predicted patterns of behaviour. The system was initiated to a significant extent by the Motorola organisation in the 1970s and 1980s.

The current state of affairs with a so-called total quality management system (TQMS) has evolved from a complex process of industrial self-auditing and *self-inspection* during manufacturing [Oakland (2000); Rahman and Bullock (2005); Samson and Terziovski (1999); Black and Porter (1996)]. TQMS involves *quality assurance* (basically quality control, validation and document control [Chzanowski (2006)]), *good overall practice* (GXP) and appropriate *planning or infrastructure*. A sketch of this evolution is provided in Figure 3.2. Here, it is clear that some considerable time ago, say for example in the 1950s, the depth of scrutiny was driven by routine line inspection and a crude attempt at inspecting quality into a end-product. This is never the case, quality can never be inspected into a product, after all testing is merely testing and there is some probability of catching non-conformance just as there is of never catching it based on the number of samples from the lot that are inspected and the criteria set. TQMS uses a notion of culture, communication and commitment to a process or 'expectation' of the customer via a quality system (tools, systems and teams) that culminates in a suitable finished product The system needs tools and techniques to more fully describe the strengths and weaknesses of a process and this is done via flowcharts, check sheets, Pareto analysis, cause and effect diagrams (Ishikawa; fishbone) and control charts (e.g. Shewhart-type charts) [Kolarik (1995); Snee (1990); Snee (1986); Buncher and Tsay (1994)].

Quality assurance was a concept borne out of inspection and progressed to what we now call TQMS. This is not the end of the line; continued rigour and demands from both industry and regulators, particularly given the globalisation of the business, mean more prophylactic steps are required to circumvent end of line mistakes. So as a rule-of-thumb if you are capable of producing non-conformity in any given process you need a quality system to prevent this occurring. This has been encapsulated in the expression *right-first-time*, where the manufacturer puts 'all' steps in place to avoid mistakes and process non-conformance. The right-first-time approach relates to the M2I model of innovative practice.

A quality system is never quite as simple as explained above. Yet at the same time it is a good deal simpler when explained diagrammatically. Figure 3.1 shows how the customer sits at the centre of the quality system activities undertaken by the organisation. The customer, meaning the patient in terms of pharmaceuticals, is served by three bodies (special product mapping, services and science) and notions of the product's inherent quality. The services are provided by industry and the most effective up-to-the-minute scientific know-how captured by researchers and scientists, engineers, clinicians, regulators and other trade professionals. These three interfaces are then regulated by factors that dictate how easily they can be disseminated to the customer via the manufactured goods. In this case we return swiftly to the notion of quality assurance.

QA practices involve all aspects of manufacturing and even some that might not be expected at first glance. The prime directive of a QA department is to push for practices in place which produce the highest purity, consistency and quality (PCQ). This can be reconfigured to mean compliance, safety and suitability

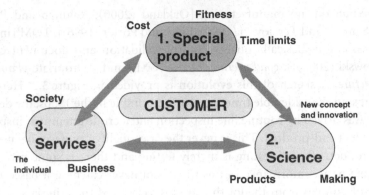

The influence of the customer is felt by 1–3

Figure 3.1 Interfacing between operating quality systems

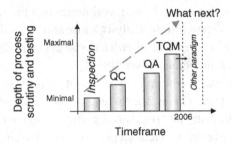

Figure 3.2 The evolutionary progression in sophistication of quality inspection

as represented in Figure 3.3. QA's 'arms' are long and do not merely relate to process analytical technology (PAT) or analytical practices such as quality control for standard products; they clearly relate to microbiological testing and are important for some high risk products such as biopharmaceuticals (see later). They are integral to every stage of production [Cundell (2004); Walsh and Murphy (1999); FDA (1996)] and standard operating procedures for undertaking practices [Chzanowski (2006)].

3.1 The quality gurus and models for assurance

In the 1950s a number of economists and mathematical scientists were given the task of assisting in the reconstruction of Japanese industrial capability and its economy. In this fertile breeding ground a new brand of business science was borne. The basis of corporate culture, efficiency and healthiness was in part assessed by the number of faulty goods produced and the manufacturing environment that produced these faults. The findings of this business culture have to date been applied to areas as diverse as automobile manufacture,

Regulatory compliance and process concerns

Compliance

QA and GMP issues

Licence held by
manufacturer
        Product 'quality'

Testing + strategies

Safety

Microbiology issues
1. Sterility, HACCP
2. Freedom from pathogens
3. End product quality
   Degradation, deterioration,
   deleterious toxins/by-products

Suitability

Product issues
1. Raw materials
2. Processing e.g. sterilisation
3. Finished goods
4. Nature of contract manufacture

Applicable to: pharmaceuticals, food (supplements) biomaterials, implants, devices, culture media, contract services

(a)

Pharmaceutical manufacture

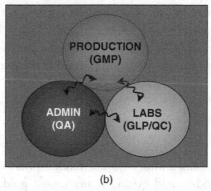

(b)

Figure 3.3 Crucial interactions and elements of successful pharmaceutical manufacture. Part (a) shows QA practices; part (b) indicates the relationship between the three key elements of medicine production

microelectronics, food and consumer goods production and pharmaceuticals. The field of initial improvement was initiated by statisticians such as Walter A. Shewhart, referred to as the father of statistical process control, and many others. Their contribution belongs in other sections devoted to control charts and the validation cycle. Notably, there are three prominent experts in terms of 'quality systems' who touch everything from research to manufacturing [Sharp (2000)], these are:

1. W. Edwards Deming:

Deming's statistical consultancy work can be said to be some of the most influential of the last century and can be summarised in terms of his expectation of product *conformity*; that is to say, the proportion of failing (non-conforming) units. It is therefore important how you construct the notion of what is acceptable and what is not. His emphasis was a notion of quality based on meeting the needs of the customer; illustrated by two of his more-often cited quotes:

- '. . . quality is satisfying the customer, not merely to meet his [*sic*] expectations, but to exceed them . . .'

- '. . . the customer is king.'

2. Joseph Juran:

Was among other things an engineer; his notion of a process and thus end product, which is one of *fitness-for-purpose*, has remained one prominent in engineering and adopted within areas of process monitoring. He promoted the idea of *quality circles* and is best remembered for significant broad-sweeping contributions to management theory. He is also noted for considerations of Pareto's 80/20 Law where the vast majority (80 per cent) of consequences result form a small number of causes (20 per cent). This theory has been expounded widely to a number of fields. In 1951 he published a significant text, his *Quality Control Handbook*. His ideas are best summarised in the quote:

- '. . . quality is customer satisfaction or fitness for use . . .'

3. Philip Crosby:

An industrialist, essayist and author who had major inputs to management theory, as we understand it today, and quality management guidelines. He is famous amongst other things for his four points of concern in practice: conformance, prevention, performance standards and the real cost of quality, that led him to consider the only appropriate vision of manufacturing was one of a *zero-tolerance approach*. In 1979 Crosby published his first business book, *Quality Is Free*. It was his firm belief that establishment of an appropriate quality system would effect savings in the long-term.

- '. . . quality is free . . .'
- '. . . quality is conformance to requirements . . .'

Other quality management science and significant TQMS innovators include Kaoru Ishikawa (fishbone cause and effect diagram), Armand V. Feigenbaum,

Shingeo Shingo (Kaizen philosophy), Genichi Taguchi (process capability, Taguchi's loss function – incremental deterioration when too high (all materials fail!) or too low (everything passes) specification limits are set), W.A. Shewhart and Vilfredo Pareto.

Deming, Juran and Crosby *et al.* were and are not the only experts in the area of quality but their postulates have formed the basis of some generic rules for manufacturing to be obeyed for the most effective manufacturing environment; these include:

- *Customer focus*, seeing the drug recipient as the external customer

- Process *relevancy* to all

- To strive for 'obtainable' *goals*

- Fostering of *employee co-operation* and innovation, including suggestion-box schemes and 'incentivisation'

- A project team that can facilitate *problem-solving*

- Proper *training and education* of employees

- A rigid high *standards policy*

- Appropriate corporate and departmental *strategies* and planning

- A mandatory accountable *guidance and leadership* structure

- The absolute necessity for *stringent valid assessment*

- The process must be furnished with appropriate investment and attributed reasonable *resources*

- To *act on feedback* from the process that is both positive and negative.

Making sure a process is right-first-time, relates to a M2I (MII) model and this realistically means '... quality cannot simply be built into a poor product...' or even simply to premises, equipment or solely raw material and to consequently have the expectation of success. Any process and its contributing elements should be constantly surveyed to identify areas of potential improvement. This is best described for any set of process variables as in Table 3.1.

Table 3.1 Possible outcomes for using set conditions during the manufacture of medicinal products

Starting stage	Interventions	Concluding stage
Good quality	→	Good, mediocre or poor quality (depending on the extent of QA activities)
Poor quality	→	Poor quality ONLY!

Table 3.1 applies equally to raw materials, excipients, services, processes, sampling methods, process analytical technologies and the active pharmaceutical ingredient and in fact takes on most of the postulates identified above from the quality gurus.

3.2 A cycle of continual improvement

A *quality circle* is a group of local experts and interested parties with particular skills who meet with the sole aim of improving the working environment. Major areas of concern are always the manufacturing process, standards and regulatory conformance, improving health and safety, improving product design, and improvement in testing methodologies based on scientific, technological and clinical expertise. Quality circles work best when there is appropriate team coherency, communication and long-term commitment and this creates a purpose and sense of continuity. The circle can operate using fixed team input from project to project or can specifically tailor the team to the individual needs of each project. Figure 3.3a (appropriate engagement) and more prominently, Figure 3.3b (task amalgamation) show the three most significant components of the team involved in quality improvement; these are production, administration and laboratories, meaning QC and R&D. Effective communication between the investors in this group can result in an improvement over and above those routine improvements and failing seen (Figure 3.4). Routine improvements are achieved by low financial input; a quality improvement group achieves a sea change and ultimately lower investiture by initially resourcing a significant effort into a marked change. Routine quality control then maintains this more effective process.

3.3 Management structure and a functioning department

An efficient quality management results from the correct interfacing of quality control, quality assurance and quality improvement initiatives. It is only through acting on feedback from those involved in the pharmaceutical product supply chain that appropriate improvement can be made. A real example of a TQMS structure, displaying common inter-relationships and essential departmental functions is shown in Figure 3.5. TQM uses a climate of expertise, communication and commitment to a process to present an overarching guide.

The centre of co-ordinated activities is the qualified person unifying ideas, regulatory guidance and research via a technical service department, via a management structure to quality assurance and its five key functions. Method and operating procedural guidance and its transfer must be optimised to give the best organisational efficiency. The schematic shows in a compact form the essential attributes of any good pharmaceutical quality system. Arrows in the

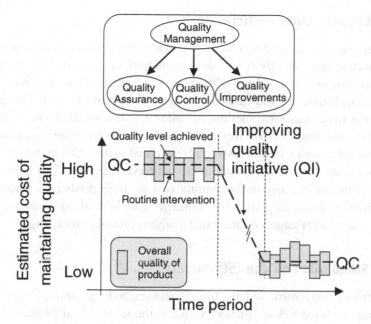

Figure 3.4 Improvement in organisational performance by routine QC and incorporation of a quality improvement initiative or process. The schematic illustrates a 'right way' to achieve systems management

Figure 3.5 The structure of one functioning set-up of a total quality management system (TQMS). Arrows indicate points of communication orchestrated by the quality management team and the qualified person

figure represent channels of communication and record keeping and in one sense these represent the weakest link in the standard process. It is for this reason that an 'accountable' person, nominally the qualified person, with appropriate skills to make a firm judgement, must rate and grade activities undertaken across the compliance, technical department and QA areas.

3.3.1 Organisational efficiency

The organisation and conformance to key elements such as appropriate levels of communication and thorough documentation to assist in QA are key to any successful outcome. The qualified person (QP) plays a vital role in the organisational structure and efficiency of any collaborative project. This means, in addition to conventional methods of information transfer, knowledge must be passed 'over-the-wall' via numerous other means to ensure repeated but structured information finds its way to all concerned parties. The 'better processes' always take into account a linear format wherever possible. This principle is certainly true, for example in the manufacture of sterile medicines, since cross-contamination could be disastrous. Linearity avoids duplication and mix up between the various stages of data transfer within product processing.

3.3.2 Standards and the ISO models

An effective management system, appropriate research QA and QC are all fully reliant on the 'yardstick of quality', which is the set standard [Kolarik (1995); MHRA (2002); Poe (2003); ISO (2000)]. In most cases this is organised through local standards, such as the European Union and EN (European norm) or British and German standards, BS and DIN, respectively. A more universally applicable standard is seen via the international standards organisation (ISO) model (Figure 3.6). Standards are set that relate to laboratory best-graded glassware, surface standards, fittings and fixtures, operating conditions and premises, statistical evaluation and sampling and even organisational structures.

United States Code of Federal Regulations (CFR) title 21 consisting of eight volumes and various sub-sections contains all regulations issued under the Federal Food, Drug and Cosmetic Act, which pertain to quality in traded commodities.

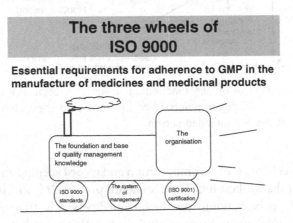

Figure 3.6 The three fundamentals of the quality system – ISO9000

The volumes are updated each year in accordance with the FDA [Moritz (2005)], federal agencies and the Drug Enforcement Administration [DEA (2006)]. Problems associated with manufactured medicines, including product recall (class I – serious to class III – limited) can be reported to the FDA/USP Drug Product Problem Reporting (DPPR) Programme which is designed to monitor cGMP and prevent defectives reaching the consumer.

Some general and more specific examples of ISO, EU (and UK, BS) standards relevant to pharmaceutical manufacture and clinical trials [Case (2006); Moritz (2005); MHRA (2002)] are given below:

- ISO-9000 Quality management systems – fundamentals and vocabulary (QA/ quality improvement)

- ISO-9001 Quality management systems – requirements (sectional roles and duties)

- ISO-9002 (BS5750 part 2; EN29002) Quality systems part 2: specifications for production and installation

- ISO-9003 (BS5730; EN29003) Quality system specifications for final inspections

- ISO 9004 Quality management systems – guidelines for performance improvement

- ISO-10012-1 (BS5781 part 1) Quality assurance: requirements for measuring equipment

- ISO-2859-3 (BS6001) Sampling procedures for inspection by attributes – statistical sampling plans (hypothesis tests). Supplement 1: sampling plans indexed to limiting quality (operating characteristic curves)

- ISO-3534 part 1 (BS5532) Statistical terminology and definitions

- ISO 15195: 2003 Laboratory medicine – requirements for reference measurement laboratories

- ISO/IEC Guide 21-1:2005 – regional or national adoption of international standards and other international deliverables – part 1: adoption of international standards

- ISO-ICS 55 Packaging and distribution of goods

- BS5703 part 3 – data analysis and quality control

- EU Directive 2001/20/EC – 'clinical trials'; Article 13: qualified person's batch certification

- EU Directive 2003/94/EC – 'GMP directive'

Figure 3.7 Simplification of the International Standardization Organization (ISO) systems relevant to the quality management of industrial-scale manufacture of medicines

- Schedules 1 and 2 of the Medicines for Human Use (Clinical Trials) Regulations (UK), 2004

- CPMP ICH/135/95 – ICH Topic (E6, section 6.4.9) (R1) – Guideline for Good Clinical Practice (2002)

There are ISO, EN and UK standards for a whole range of processes, products and activities (Figure 3.7). In the pharmaceutical context the primary ISO (governing) standards are ISO 9000–9004.

A famous quote from Deming seems appropriate at this juncture: 'Cheaper is not always better BUT better is *always* cheaper.' This holds true across many boundaries but particularly with regard to setting the standards by which discriminatory tests and judgements will be made, such as those involved in analytical and clinical validation, where financial 'short-cuts' can be a source of profound error or bias (μ_1, described in Section 6.2).

3.3.3 Kaizen, quality circles and the quality spiral

Kaizen Teian (*Kaizen*) is a business philosophy that comes from the mechanical engineering, automotive and microelectronics industries based on quality circles, continual improvement and full employee participation. It encompasses a series of step improvements in a process that might arise from the troubleshooting efforts of a quality circle. The expertise encompassed in a quality circle team traverses the unit operations (discrete activities) routinely found within any manufacturing process. In the case of pharmaceutical manufacturing this would involve the QP (an expert in their own right), higher management member, production expert, supply chain expert, pharmacist, process engineer, pharmacologist, clinician, chemist or analyst, microbiologist, software engineer, marketeers and sales team, business manager, product regulation and law expert. Quality circle team members usually act on a voluntary basis so as to avoid elements of coercion or payment for actions and have the directive that work be undertaken in the best interest for the customer; this is best done through *contributor enabling*.

The quality spiral is intimately connected with the activities of the quality circle. Figure 2.1 in the simplest form shows its basic format, based around a drive *for continual improvement* and an iterative involved cycle of internal audit and review (see also Figures 3.8 and particularly 3.9). The notion of a quality spiral is generic to most processes and so the elements that constitute each tour of the cycle vary. These may include process analytical technologies, validation and regulatory compliance during manufacture but equally might involve method selection, system suitability indices and chemometric capability as part of analytical aspects of QC or HTS. The reliable and valuable findings from each cycle are retained and promoted to those used as standards in the next (more refined next approach) cycle of activities; those of little or deleterious value are discarded. Thus each cycle links to a better cycle until the objectives are fulfilled.

An iterative approach as described in the quality spiral ensures a process of continual improvement. Other approaches used in modern-day pharmaceutical practice are based on tactical philosophies to lower general wastage:

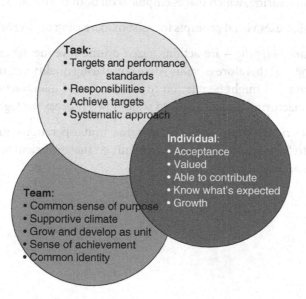

Figure 3.8 The characteristic qualities required from a successful development team

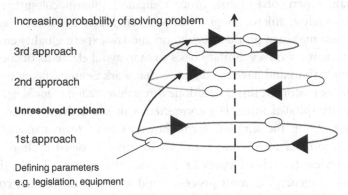

Figure 3.9 Representational portrayal of the quality improvement (quality control) spiral and its influence on problem-solving

- *Hoshin* – a policy deployment related to breakthrough and business fundamentals.

- *Poka-yoke* – strives for defect-free products by backing-up mistakes in operations, routinely (examples: limit switches, specialised fixtures, colour coding).

- *Just-in-time* – Japanese concept from around the 1950s; this is a planned manufacture of the exact quantities as needed without waste and excess (inventory management). This is now considered to be an integral part of '*lean*' manufacturing, which places emphasis on both quality and productivity.

- *Kanban* – that uses visual prompts to assist in increasing or decreasing activity.

- *Value Stream Mapping* – are actions based on the revenue generation of the organisation and therefore of value when comparing quality versus productivity/profit and that might be relevant to some forms of manufacturing such as 'lean' manufacturing that attempts to scale-down process wastage.

- *Total Production Maintenance* – divides the unit operations involved in a process into 'bite-size' modular components so that sufficient weighting can be given to each action.

SECTION B

Setting Process Bounderies

4
Validation

Validation is a planned series of interactive tests and inspection designed to describe and reduce uncertainty in an important process; it can span analytical science, microbiology, and testing of software such as laboratory information management systems (LIMS) [Wagner (2006); Powell-Evans (2002); Friedli *et al.* (1998)]. Appropriate system scrutiny is made possible by following Shewhart's cycle of development. The four components of this system represented in Figure 4.1 are all strategic steps followed to get the best results from a validation exercise. Validation activities which show that a process or product works to a suitable standard are based on:

- Studies (tests or qualifications) showing performance (and intrinsic behaviour) characteristics.

- Proficiency testing which uses the notion of systematic parallel testing of two systems.

- Internal auditing (sample testing) of a process or product:

- 100 per cent of all material made is evaluated – but this is both destructive and costly

- Statistical – uses mathematical models to extrapolate the outcome from a limited number of evaluated pieces

- *Ad hoc* – where a pre-set number of pieces are evaluated to indicate a quality level.

- External auditing – of suppliers and service providers.

- Challenge testing, where the most extreme 'worst case' is applied. If the product passes it is likely to pass all routine conditions.

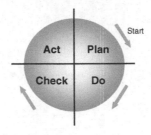

Figure 4.1 Shewhart's PDCA cycle of general validation activities and process development

- Operation characteristics of a process or product, given the limits of apparatus and personnel.

- Failure acceptance-improvement culture (so-called quality spiral or Kaizen approaches are applied to improve on the existing performance).

The PDCA process is initiated by an appropriate planning stage (plan), followed by implementation of the plan (do), this is followed by an interrogation and confirmation step (check) based against standards already pre-determined; based on the outcome of this stage the results and qualification data are acted on (act). If the validation fails or is unsatisfactory another cycle may be initiated.

The compliance of a process to the standards set-up by the organisation can be calibrated against existing plans or regulations. This often happens as a consequence of enforcement by inspection that involves:

- Regulators such as the MHRA, FDA, Irish Medicines Board, etc.

- Cross-referencing to the strategic plan of the process/product as detailed in the validation master plan (VMP) that contains the most important of protocols, defining: responsibilities, nature of testing, items concerned, details of operation.

Validation of a process can be undertaken in various modes depending on the amount of supporting information required, limitations of inspection and the risk deemed acceptable by the organisation. The latter should be as low as is possible to achieve. Key types of validation are thus:

- Prospective validation mode – where testing is performed prior to routine manufacturing

- Retrospective validation – review of historical records (not normally advised) and return to 'plug' spaces in knowledge concerning a product

- Re-validation – involves a repeat to provide assurance of lack of effect in changes incorporated into a process no matter how subtle.

The basic steps of any generic validation must take into account the following milestones before the process or product can be considered to have been validated. The so-called validation lifecycle should consider the following:

- Commissioning of new apparatus and equipment based on the user-requirement specification (URS) which details the needs and 'wants' of a process to obtain the desired goals. The URS lists all the key desirable attributes of a new piece of equipment for example.

- Installation of commissioned new equipment and the (a) installation qualification (IQ) and (b) basic operation qualification (OQ) which test the general suitability of the process/product *in situ*. Failures in the IQ/OQ [Greaves (2006)] lead to re-commissioning and then re-installation tests.

Product validation is needed for all activities that impact on product quality, form and general suitability in the present and near future. In the case of generic testing of a pharmaceutical process or analytical methodology for use in the lab the validation life cycle progresses as follows:

- Design qualification (DQ) – the matching of new items/facility against the URS by review of the design. This is sometimes further expanded for an equipment qualification (EQ) and is routinely undertaken when new analytical apparatus is commissioned.

- Installation qualification (IQ) – the performance/document checks that are minimally essential for basic operation and that might be relevant to future calibration.

- Operation qualification (OQ) – provides the 'critical' tests (and repeats) to verify system performance.

- Performance (making the product) qualification (PQ) – the documented demonstration of 'processing' for routine specific operation for (a) product compliance and (b) OQ tests. It may also provide information say for setting the limits for cleaning validations where medicines are concerned or in a process analysis context where cross-contamination is possible.

These four basic test steps form the basis of a generalised validation plan.
Process validation actually has a distinct meaning from 'validation' as it refers to the documented demonstration of a process and its suitability that is based on *three* consecutive batches of product. In this context the 'proper' validation of a process, called the comprehensive validation pack (CVP) consists of: process validation based on three batches, appropriate method validation [Munden *et al.* (2002); Cledera-Castro *et al.* (2006)] for all the methods involved in the making of a suitable end product, and these have to checked against specifications for official

(professional body) compendia and international conference on harmonisation (ICH) compatibility. Basic pharmaceutical dosage form validation should take into account the analytical chemistry method of analysis (in terms of analyte recovery), most appropriate form of assay, representative physical sampling, instrument or room dedication to lower the risk of cross contamination, and the use of suitable and method-compatible cleaning products.

Process validation may also have to consider computerised system validation for all programmable logic controllers (PLCs) that run analytical, process and on-line process monitoring equipment. This is essential because in today's environment much of the interpretive assessment and automation is undertaken by PLCs. The relative importance of PLCs was highlighted at the end of the last millennium as manufacturers and equipment suppliers struggled to guarantee that their controlling software would register the change of the year 1999 to the year 2000, thereby invalidating output from equipment. In addition to software itself, standard operating procedures (SOPs) require validation to confirm appropriate methodologies are being used throughout a process. Gap analysis is often undertaken to highlight significant shortfalls in proper validation that is often based on assumptions (e.g. for vendors and contractors) that external activities have been undertaken fully and in compliance with site standards. All suitable process validations and particularly those pertaining to cGMP-based pharmaceutical manufacture require a summary report to indicate the uncertainty or confidence in the testing undertaken and its validity [Mollah (2004); Slater (1999); FDA (2001); FDA (2003); Moritz (2005)].

Validation of a process is necessary because no manufacturer of medical or pharmaceutical therapies or devices wants to produce a final product that is injurious to health, and risk losing the organisation *product license*. However, in general terms the real reason to undertake comprehensive validation is to increase understanding of the product/process and to perform an assessment of potential risk (BS, EN, ISO – compliance, see Section 3.3.2). This leads to the smooth and efficient running of routine manufacturing that reduces regulatory non-compliance and ultimately leads to less product testing, less wastage and less of a need for repeat procedures to be undertaken (redeployment costs). Implementing these higher quality standards needs appropriate resources because it requires investment in proper certification and training of relevant personnel. To this end there must be an efficient department and critical mass of personnel to permit such costs and resources. Validations as such are usually undertaken in two forms (a) top-down where the final product and then components are assessed, and (b) bottom-up (better) where the individual pieces that make up the bigger picture, and the picture itself in its entirety are evaluated (see later for more detail of practical use). Both these approaches are useful but unless required it is generally advised to use the more safeguarded bottom–up approach to process validation (Figure 4.2).

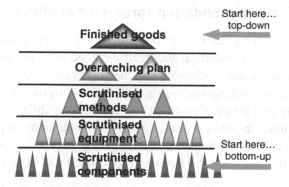

Figure 4.2 Validation made simple: the top-down and bottom-up methodological approaches

The set conditions for any qualification require that the test be consistently performed under specific restrictions. For diagnostic purposes such validation makes use of the following 'fitness-for-purpose' aids:

- System suitability tests, such as accuracy

- Challenge testing, where the equipment is stressed more than would be customary

- Method transfer from recent research and development with best practices involved

- Failure-acceptance-improvement culture (MII, quality control spiral).

These all serve to increase the evolution to the best method in current practice and consequently validation procedures should not be considered static but constantly upgraded.

4.1 Process and manufacturing validation activities

Process validation (PV) centres on confirming that a process sits within the specification listed as part of individual qualification (tests). In general terms, process validation refers to all the independent activities that are part of a 'process'. The process normally refers to an action with a definitive product or outcome and an *internal* or *external customer* of the organisation. As such, process validation usually means manufactured goods but could in one example, relate to activities or services, for example within a clinical environment such as a clinical trial of a new drug product or software beta-testing for a programmable logic controller (PLC) [Powell-Evans (2002); Friedli *et al.* (1998)].

4.1.1 Manufacturing validation (processed product)

This represents perhaps the biggest portion of the entire validation cycle of any manufactured pharmaceutical goods, although no more important than any other. Other elements such as cleaning and analytical validation are amalgamated with manufacturing validation as in-process (IPC) or on-process control (OPC). The overarching goal is one of ensuring end-product suitability (against quality indices) by fragmenting the process into modules with an appropriate consideration of risk and non-compliance to established standards that come from initial method transfer from the R&D and local institutional technical services environment.

Matching of practice to established standards is made possible by reference to suitable FDA (internationally relevant) manufacturing information [FDA and GSK (2006)], UK regulation [MHRA and Roche (2006)] such as supplied by the MHRA [Mollah (2004); Norris and Baker (2003)] or the guidelines supplied by the regulatory body in the country concerned. As such the essential considerations of any validation of manufacturing should be:

- Whether to establish a new, or alternatively use a revamped or existing protocol (method)?

- Establish if there is sufficient current in-depth understanding, certainly an issue with a new chemical entity (NCE)

- The need to follow and establish an environment of cGMP [MHRA (2002)]

- The skills and resources of personnel and organisation, respectively

- The site/building/equipment limitations

- Aspects associated with packaging/storage/handling of the product

- The sophistication and complexity of routine production

- The detail of testing and lab control

- Provision of a suitable 'audit trail' in terms of detailed records/reports.

Additional aspects of higher-end quality in a manufacturing (process of making the product itself) validation should also consider:

- The depth of experience among the key personnel and qualified persons that relate to the product or the product type.

- Possibilities for use of a strategy that relates to generic validation (simple products).

- The likelihood for consistency of manufacturing and the consequences of inconsistency. This is particularly relevant to high-risk parenteral-type products (see later, Section 5.2.1).

- Exact and clear (unambiguous) definition of *all* the steps in a process.

- Useful FDA/MHRA (MCA) guidelines that might be used directly or modified in the case of a NCE.

- Use of pilot trials to give an indication of the point of 'weakness' in a particular process where more relative attention should be focussed.

- Wherever possible making use of pre-established validated methodologies in order to reduce the burden and workload; such methods may be obtained from referential guideline organisations that include:
  - DIN (German Institute for (Norms) Standards)
  - UKAS (UK Accreditation Service)
  - NIST (National Institute for Science and Technology)
  - NBS (National Bureau of Standards)
  - BSI (British Standards Institute)
  - ASTM (American Society for the Testing of Materials)
  - ISO (International Organization for Standardization, Switzerland).

All attempts at validation must be initiated by reference to a validation plan (VP) that details all the information required to establish a correct validation. It usually in a most basic form consists of at least an introduction, process-flow diagram showing the interconnection of elements to be considered, and provides details of key tests as described by essential standard operating procedures (SOPs) and manuals. The VP may also suggest the most essential results of any qualifications and guides to decision making. The validation plan is always concluded by a validation report that represents conclusions to the testing undertaken and its status in terms of pass or fail and the need for further testing based on any uncertainties.

Validation is undertaken in an attempt (a) to provide evidential proof of process compliance but also (b) to provide the ancillary element of an assessment of risk (Figure 4.3). This is clearly the case with products where microbiological quality and freedom from pathogens might be called into question. As the process is progressed and a greater investment of time and money takes place this risk of product non-compliance (failure) becomes more significant.

During routine manufacturing of pharmaceutical products the principal concerns would be:

- The drug substance – purity and freedom from degradation products

- Product stability (compositional change) and impact this might have on bioavailability

- The correct status of the certificate of analysis (CoA) and the results of laboratory testing of the product

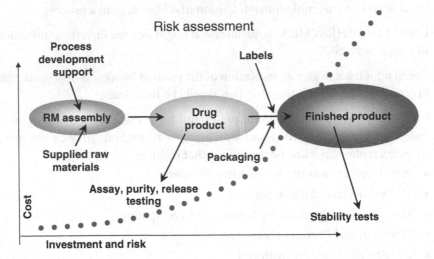

Figure 4.3 Organisational investment and risk represented as a function of the stage in a routine process validation

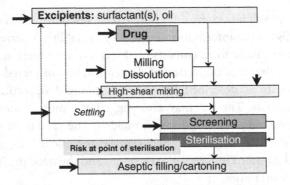

Figure 4.4 Generalised manufacture of sub-micron-sized conventional parenteral emulsion product. Both the particle size and product sterility are principle concerns that relate to patient safety

- The integrity of the drug product at all stages during and after assembly (Figure 4.4).

During processing that might involve heating (as indicated during sterilisation), drying, use of solvents or irradiation, ethylene oxide and mixing of components there is the additional risk of the inclusion of drug impurities (and toxic degradation components) and of a compositional change in the product (Figure 4.5). On another level there is a considerable risk of incorporation of foreign matter that might signal an adulteration of the product, which may not

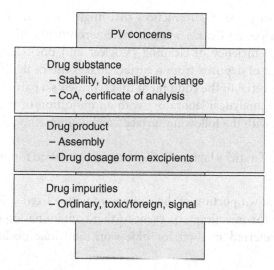

Figure 4.5 The routine concerns of process validation from the good manufacturing practice (GMP) perspective

necessarily make the product unsafe but might reduce the quality and acceptability to the customer, such as soot particle in an ampoule-based medicine, or discolouration of a coated tablet.

4.1.2 Cleaning validation

Taken as distinct from manufacturing validation although obviously related, considered by many to be a part of manufacturing validation, and necessary before any commencement of routine production of any pharmaceutical product where cross-contamination is possible and the risk of that contamination could adversely affect the product superficially, or significantly risk product non-conformance. In this case we are always asked 'what is effective cleaning [Verghese (2003); LeBlanc (2000); Munden *et al.* (2002)]?' For any given industrial manufacture of drug products cleanliness is customarily assessed by using sampling of the surface, either directly (swabbing a known area of the surface) or indirectly (placebo drug and carry-over contamination). Typically, limits are based on some 'rule-of-thumb' of experience-based practical guidelines [Sarker (2004)], such as surfaces being 'visibly' clean, or when a sampled portion contains 0.1 per cent of the customary dosage strength of the drug, or an arbitrary limit, for example of 1 ppm (μg/cm^3 or μg/g) in a specified volume or from a specified surface area. Limits must be carefully considered to mean that cross-contamination is permissible at the level recorded. For sense to be made from sampling (and thus cleaning validation) it is therefore necessary to clearly define all operating conditions. In modern-day production use of specific clean-in-place (CIP), wash-in-place

(WIP) or steam-in-place (SIP) practices that might be used in biotechnology, fermentation devices and other automated environments still require careful validation of the efficiency of cleaning [Vogleer and Boekx (2003)]. Routine testing of removal of deposits from a surface makes use of flat 'coupons' made from identical material to the equipment that represents a part of the apparatus. This provides the analytical laboratory with an indication of the difficulty of a cleaning practice with the following surface assessment methods:

- Direct contact of a sticky layer with the surface (often used for micro-biological assessment).

- Rinsing of a known portion of the equipment with a fixed volume of solvent (usually water-for-injection to prevent further contamination of the surface). This is the preferred method for pipe-work and inaccessible locations of equipment.

- The most common technique uses cotton (poor recovery) or fully dissolving alginate swabs to recover the surface contamination from a fixed area of the equipment surface. The surface scrutinised is usually an area of 10 cm by 10 cm and located in the more problematic regions of the apparatus.

Alternative methods for assessment of the contamination burden in the air may be based on:

- Air-impingers and filters which draw in air, and then subsequent dissolution and testing

- Settle plates which allow the 'contaminating particulates' in the air to fall on a surface over a fixed time period to provide an indication of concentration.

In general, the chemical assessment of drug and excipient contamination on production surfaces or in the air within a production suite is based on the following ubiquitously used technologies:

- Wet chemistry assay such as titration.

- Chromatographic assessment such as by gas or high performance liquid methods.

- Spectroscopic methods such as total organic carbon (TOC) or ultra-violet/visible spectroscopic assay and near infrared (NIR) evaluation. In some cases such as the NIR technology it may be possible to use a reflective mode to assess the surface contamination directly using fibre-optic technologies.

- Surface contamination involving charged or ionised molecules can be assessed using pH or other electrodes.

Sampling considerations [Ahuja and Scypinski (2001)] when recovering surface contamination for lab assessment and commencement of a new campaign of manufacture are to use the 'worst case scenario' model that might involve inaccessible, pyrolysed products or desiccated material. SOPs for cleaning validations should record the exact volumes and conditions of physical sampling in addition to detailing the location of the sampling from the surface, and permit sample holding for less than a day before release. Limits should include a safety margin to account for random variations in the surface contamination from batch-to-batch of produced material.

Establishment of extreme limits for surface contamination can prove useful as a basis for commencing or stopping production. Acceptance criteria for cleanliness may be accounted for from the following ratios:

$$C_{max} = (d_t \times b_1 \times F)/d_{next}, \tag{4.1}$$
$$C_0 = (LD_{50} \times F), \tag{4.2}$$
$$ADI = (C_0 \times M \times F), \tag{4.3}$$
$$C_{max} = (ADI \times b_2)/d_{next}, \tag{4.4}$$

where: C_{max} = max allowed carry-over, d_t = therapeutic dose, d_{next} = largest daily dose of following manufactured product, b_1 = batch size, b_2 = smallest batch of any subsequent product, F = safety factor, C_0 = no observed effect level (NOEL), LD_{50} = lethal dose for 50 per cent of animals tested by intravenous route, ADI = acceptable daily intake, M = average adult weight. Many values can be obtained from literature or clinical trial and toxicology study data.

Equation (4.1) is perhaps the most universally applied form (others are used, for instance, where limited information is available) for surface evaluation and in this case a safety factor (Table 4.1) is needed to scale the maximum allowed carry-over concentration. With respect to safety concerns certain substances are highlighted specifically and these include penicillins, cephalosporins, potent steroids, cytotoxics, allergenics and endotoxins, which have limits less than the limit of detection (LOD) for the substance concerned by the best analytical method available.

Table 4.1 Rule-of-thumb safety factors (F) for various pharmaceutical products (applicable to Equations (4.1) and (4.3))

Normal daily dosage permissible as carry-over	Dosage form
10–1 %	Topicals: creams, lotions, liquids^
1–0.1 %	Oral: tablets, capsules, liquids*
0.1–0.01 %	Parenterals: injections and ophthalmic liquids#
0.01–0.001 %	Research candidate drugs/NCEs

May include: ^liniment; *syrups, extracts and tinctures, draughts, linctuses; #irrigant [Ansel et al. (1999), 244–262].

4.1.3 Software and PLCs

Again as with cleaning validation, software validation of programmable logic controllers (PLCs) could be taken to be a part of general manufacturing validation, since today it is implicit in any complex process. It is separated in many cases for reasons of practicality and extent and to facilitate parallel tests to be conducted and with a notion that software and 'informatics' science is constantly being refined on the time-scale of months. Their influence in modern day manufacture is ubiquitous. Some problems arise, such as the not so distant 'year 2000' compliance issues, and changes to the law, that were not envisaged at the initial validation of the process-controlling software [Advanstar Communications (2003); Norris and Baker (2003); Deshpande (1998); Loftus and Nash (1984); Case (2006)].

4.1.4 HACCP

Any process involving multiple unit operations needs a hazard analysis of critical control points (HACCP) and a hazard and operability (HAZOP) study before commencement. HACCP is necessary to identify and isolate areas of inherent augmented risk. Typically this might involve issues concerning sterility or freedom from microbial toxins and by-products. HAZOP is intended to comply with employee working conditions, which might be particularly important with cytotoxic or experimental uses of newly developed chemical entities or difficult working conditions such as aseptic manufacture [Mollah (2004); Selkirk (1998); FDA (2003); FDA (1996); Cundell (2004)].

4.1.5 Ready to start manufacturing?

Making sure the manufacturing environment is clean and organised before initiation of the product run reduces the chances of making mistakes and wasteful manufacture. It is mandatory before commencement of any pharmaceutical production for the production team to ask the following questions:

• Am I ready and appropriately set to start manufacture?

• Have we got the right procedures in place to avoid interrupting the run?

• Are decisions to be based on judgement (say for example of the QP) or indisputable chemical assessment?

• What are the consequences to the organisation and customer for getting an assessment of readiness wrong?

When answers are in full and to the full satisfaction of the production staff (and quality control manager) it is appropriate to start manufacture. An appreciation of the need for the quality (control) spiral or 'Kaizen' to be applied across routine production is required because this provides an additional source of best practice information in support of process validation.

4.2 Valid analytical methodologies (VAMs)

Analytical and diagnostic chemical methods also require validation and for much the same reasons as any other type of validation [Buncher and Tsay (1994); Ahuja and Scypinski (2001) Kellner *et al.* (1998); Benoliel (1999); Powell-Evans (2002); Skoog *et al.* (2000); Harris (1999)]. They are undertaken to ensure the 'fitness-for-purpose' of process-related analytical chemistry and the chemistry associated with clinical trials, new drug development, HTS and pharmacological investigation. The essential function of VAM scrutiny is to monitor the control procedures that will be used to decide whether a process is working correctly. Therefore deciding if it is entirely pertinent to test the working of the test is an essential part of using a reliable method.

Valid analytical methodologies can be obtained from a number of sources some of which may include:

- IUPAC (International Union of Pure and Applied Chemists)

- AOAC (American Organization of Analytical Chemists)

- AMC (Analytical Methods Committee) of the Royal Society of Chemistry (United Kingdom)

- Local pharmacopoeial (BP, EP, JP, USP) and compendial testing

- FDA or other regulatory bodies' specifications (for new drugs) and cautionary notices.

The quality control section must know a process well; most failings in good quality testing come from making assumption across many levels. Good practice is best summarised by the expression: 'to *ASSUME* is to make an ASS of U (you) and ME'. What this really means is not having uncertainty of method, or protocol followed, at any stage [Taverners *et al.* (2004)].

4.2.1 Model systems and GLP

Good practice in the laboratory is based on sound assessment of the materials and methodologies used in a purposeful manner. Considerations are always:

- Technique selection criteria – the best (most discriminating) method is always used

- Use of the most pure and well-defined materials and reagents (Analar grade)

- Appropriate physical sampling and the capture of a representative sample reflecting the composition of the drug product is often a point of failure in analytical assessment, as such proper sampling is essential to good practice.

The sampling of material from a lot of manufactured product makes use of an inert container and procedural activities that reduce the chances of altering the sample composition; these might typically include evaporation from the sample, temperature cycling and exposure to light. The means of taking a sample of the product for further testing and methods are illustrated in Figure 4.6.

Batch-wise sampling in addition to the methods shown in Figure 4.6 may make use of rifling where a purposely-created pipe is used to decant solid sample. Sampling of solid is subject to the problems of 'sifting' in the powder or 'creaming' or sedimentation in semi-solids and liquids; this further complicates reproducible capture. Continuous sampling that relates to liquids or gases is undertaken *in situ* but is not as widespread as sampling of solid or liquid in batch form. Roughly 75 per cent of all the errors introduced into flawed assessment of the content of a sample are made at the initial point of sampling.

The start point should contain a bulked sample (Table 4.2) taken from the top-middle-bottom and left and right of the sample lot container to avoid bias (μ_1) and unit sifting or systematic errors. Any method that recovers drug for estimation of content and product suitability must consider possible degradation of sample (e.g. as the pH is modified for a method of analysis) that may take place in solution form, and the speed and timeliness of measurement. Effective loss of the sample and what is considered to be effective recovery of sample (not less than 80 per cent; see the section on sample recovery, below) is an important consideration in modelling of content of uniformity and also in cleaning validation.

An essential part of GLP and the making of sound measurements [Webster (1995)] is the careful management of records, ongoing control and rigorous

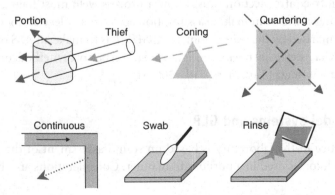

Figure 4.6 Laboratory sampling methods and physical recovery of a test sample

Table 4.2 Taking analytical test samples from lot manufacturing of pharmaceuticals using bulking and subdivision technique to ensure sample consistency

Protocol for permitting analytical measurements on solid, liquid, powder-based samples		
Stage	**Start**	**End**
Step 1 - bulking	'x' units of final product lot	–
Step 2	Combine units	Produce sample unit composite
Step 3	Sub-divide sample unit composite	Produce sample lots A, B, . . . , n
Step 4 – lots for QC tests	Sub-divide each of sample lots as appropriate for number of individual tests required	Produce sample sub-lots A_A, A_B, A_C, B_A, B_B, B_C, etc.
Step 5 – tests	Parallel tests, stability sample (stored), reference sample (untouched)	

regular instrument calibration. Other crucial aspects of GLP include safety in the laboratory, appropriate tidiness and cleanliness, adherence to the SOP, extensive documentation, on-going control and assessment of practice and the disciplined organisation of the analyst.

Sample recovery

This is vital for valid estimates of cleanliness or where matrix effects (poor dissolution or recovery of analyte) might be observed, hence appropriate calibration methodology is required. Sample recovery is an essential part of physical sampling and of sound capture of a representative, uniform analytical sample. It is not always entirely clear what represents good recovery. However, coupon trials (sample recovery calibration) can be used to establish difficulties in sample recovery associated with issues such as drug or excipient solubility. The coupon approach makes use of similar surface fouling, abrasion and soaking that might take place in the experiment. This is very useful for CIP–WIP system modelling where coloured contamination (such as harmless, yellow staining riboflavin) may be added to a surface to test sample removal.

Key analytical methods

Many analytical methods are used to establish product suitability in the long-term. For these reasons laboratories often use accelerated testing equipment that effectively speeds up a process so that product suitability can be assessed in a matter of hours or days rather than months and years. The use of such equipment is in getting hold of key information quickly. One type of apparatus, the Rancimat, uses intense light and oxygen to simulate the natural aging of oil-containing products so that their susceptibility to rancidity can be evaluated. Similar equipment now also exists for bacteriological assessment and the aggregation and aging of colloidal materials and coarse dispersions.

The crucial tests in the routine assessment of medicines usually involve evaluation of pharmaceutical actives and excipients and an assessment of texture and physical form (Table 4.3). This list would be incomplete without an additional reference to the microbiological status of the product. Key tests are thus:

1. Chemical assessment – assay, water content, mineral ash, impurity limit tests

2. Microbiological assessment – indicator organisms, pathogens, biochemical tests, pyrogenicity (limulus lysate amoebacyte, LAL test), most probable number (MPN), plate counts/selective and differential media and process sterility strips

3. Pharmaceutics assessment – rheology, particle size, hardness, friability, polymorphic form, dissolution profile and packaging integrity.

The hygienic and microbiological status of the product is at the 'top of the list' among indices of quality and safety (Figure 3.3a). Microbes exist in a complex number of sizes, shapes and forms. These include minute viroid particles 300 or so base pairs long, 20–300 nm virus particles, 0.3–0.8 micron microbacteria and 0.75–5 micron bacteria (rods, cocci, etc.) and spores [Hanlon (2002)]. Microbes also include yeasts and moulds that can ferment pharmaceutical products. Basic tests include those for are for gram negative rods (e.g. Psuedomonads and Coliforms with lipopolysaccharide endotoxin release) and gram positive (e.g. cocci, *Bacilliaceae, Lactobacilli* having a simpler structure) bacterial forms involving crystal violet stain and endospore content (e.g. *Clostridia, Bacillus*) using malachite green. Spores are very resistant to heat and can easily withstand temperatures of $80°C$ for 10 minutes that would inactivate most vegetative bacterial cells, and therefore represent a possible breach of hygienic status. Sterile parenteral products should have sufficient processing history to reduce the risk within the product. Typical other tests might include identification by specific stains (e.g. ATP specific, mycobacteria specific) and various types of microscopy (phase-contrast, fluorescence, dark-field, electron microscopy). The standard form of enumeration is undertaken by spread and pour plate methods of agar plate culture under specific gaseous and temperature conditions using culture media with controlled formulation. Microscopic cultivation and counting is also possible using impedance and conductivity measurements along with other automated protocols such as direct epifluorcscence technique (DEFT), which use an ATP-sensitive dye (luciferin/luciferase) to give an indication of live cell count. Familiar biochemical tests include sugar fermentation, oxidase, urease and catalase enzyme activity, gelatin hydrolysis, blood hydrolysis, citrate utilisation and pH change. Many of these tests are now found in rapid off-the-shelf bacteria identification strip tests.

Table 4.3 Evaluation of common analytical, pharmaceutics science/formulation and microbiological survey methodologies

Method	Application and use	Specific analytes/ use	Sensitivity	Linear performance (linearity)	Analyte selectivity	General technique precision and accountability	Comments
1. Analytical sciences							
Pharmaceutical titrations/wet chemistry/limit tests	Assay/ identification	Many	10^{-3} M	$1-10^{-3}$ M	Poor	Good	Universal application
Ultra-violet/ visible/fluorescence spectroscopy	Assay/ identifi- cation/QC	Conjugated structures, substituted groups and derivatised moieties	Substantially less than 10^{-6} M	To 1 M	Poor	Good	
Infrared spectroscopy	Identifi- cation/QC	Functional groups	1 mg/full cell required	0.1–5 mg	Good	Poor	Universally favoured for identity confirmation
Flame photometry	Assay	Alkali metal salts and buffer salt ions	10 parts- per-billion (ppb)	100-fold	Good	Good	Limited but useful
Atomic absorbance and fluorescence spectroscopy	Assay	Metals, metalloids, heavy metals, pro-oxidant catalysts	1 ppb	100-fold	Very good	Good	
Nuclear magnetic resonance spectroscopy	Identification	Functional groups	10 mg needed	n/a	Good	Satisfactory	Molecular modellers favour this technique

(continued overleaf)

Table 4.3 (*continued*)

Method	Application and use	Specific analytes/use	Sensitivity	Linear performance (linearity)	Analyte selectivity	General technique precision and accountability	Comments
Mass spectrometry (various forms)	Identification	All	Gram quantities	n/a	Very good	Satisfactory	
Refractive index spectrometry	Assay	Liquids	More than 1% w/v	10-fold	Poor	Good	
Optical rotary dispersion and circular dichroism	Assay	Isomers and polymers	Less than 1% w/v	10-fold	Poor	Good	
Ion selective electrodes and 'electrochemical methods'	Assay	Ions, pollutants and electrolytes	Excellent. Substantially less than 10^{-6} M	10^5-fold	Satisfactory	Good	Not generally user friendly
Thin-layer chromatography and gels-lab electrophoresis	Assay/identification/QC	Many including impurities	100 ppb	10-fold	Very good (by use of spray reagents)	Very poor	Universal in Pharmaceutical QC
Gas/liquid/headspace/supercritical fluid chromatography	Assay	Many/volatile compounds	100 ppb	10^5-fold	Very good	Good	

Table 4.3 (*continued*)

Method	Application and use	Specific analytes/ use	Sensitivity	Linear performance (linearity)	Analyte selectivity	General technique precision and accountability	Comments
High, ultra-high and fast protein liquid chromatography and capillary electrophoresis	Assay/ identification	Very many	0.1 % to ppb (depending on detector type)	10^5-fold	Very good	Good	Universal in Pharmaceutical QC
Thermal analysis (DSC, DTA)	Identification	Pastes, solids	Sub-mg quantities	10^5-fold	Poor	Good	Universal in Pharmaceutical QC and polymorph identification
2. Pharmaceutics science							
Rheology, material tests	Assay/QC	Hydrocolloids, suspensions and polymer melts	Semi-solid to 1 mPas, (water value)	Good	n/a	Good	Universal
Particle sizing and imaging	Assay/QC	Particulates	2 nm-visible particle diameters	Satisfactory	Satisfactory	Very good	Need regular shapes for reliable data
Dissolution & disintegration time	Assay/QC	Tablets, capsules, pessaries and suppositories	Satisfactory	Satisfactory	Poor	Satisfactory	Universal
Hardness and friability	Assay/QC	Tablets and solid dosage units	Poor (COV few per cent variation expected)	Poor	Poor	Poor	Universally used in process guide

Table 4.3   (*continued*)

Method	Application and use	Specific analytes/use	Sensitivity	Linear performance (linearity)	Analyte selectivity	General technique precision and accountability	Comments
3. Microbiological sciences							
Selective and differential microbial growth media	Assay/indicative micro-organisms/QC	Aqueous preparations of all formulations	Satisfactory	Satisfactory	Good	Satisfactory	Universal
Pathogens and spore-formers	Assay/QC	Aqueous preparations of all formulations	Satisfactory	Satisfactory	Good	Satisfactory	Used for sterile products and parenterals
Biochemical kit tests and enzymic diagnostics	Assay/QC	Microbe metabolite functional groups	Good	Satisfactory	Very good	Satisfactory	Used as identification aid e.g. for pathogen strains
Most probable number (MPN)/sterility tests	Assay/QC	Asepticity	Good	Satisfactory	Satisfactory	Satisfactory	Universal

n/a – not applicable. Essential tests are provided [Skoog *et al.* (1998), 299–562; Sarker (2004)].

Yeasts and moulds are more complex organisms than bacteria and some produce a filamentous network, fortunately less than 0.1 per cent of these organisms are pathogenic. However, they can be responsible for fermentation, production of fibrous structures, and production of potent toxins in some notable cases and digestion of the ingredients used to give pharmaceutical products a texture. Yeasts are typically 2–4 microns in size and ferment sugars; common examples include *Saccharomyces* spp. Gas production is detrimental because it could cause bottles and packs to burst. In some cases they are not passive and harmless and can prove to be of significant influence in immuno-compromised patients. Moulds include species such as *Mucor, Aspergillus* and *Penicillium* spp. The first two produce ergotamine-like toxins and aflatoxins or ochratoxins. These can have both short-term and long-term effects on the liver.

In terms of preservation it is important to know about minimum inhibitory concentrations for these micro-organisms. Other key product tests would include atmospheric monitoring and disinfection evaluation [Hodges (2002)]. The European Pharmacopoeia (2000) recommends testing for four key organisms. These are based on the product type, so topical and non-sterile products have limits for *Ps. Aeruginosa*, and *S. aureus* and oral and herbal products are screened for threshold levels of *E. coli, Salmonellae* and *S. aureus*. Sterile medicines should be free from pathogens and non-pathogenic microbes. *Bacillus stearothermophilus* strips and *B. subtilis* are used to monitor sterilisation by moist heat and *B. pumulus* is used to interrogate products for adequate irradiation. Nowadays, chemical heat and radiation-sensitive tapes, strips and indicators are also used. As part of any validation exercise the removal of Bacilli and Clostridial endospores and the indicator organisms used for non-sterile medicines would also be made in addition to a species of yeast and mould. One might also consider the case for removal and screening of prion proteins from suitable bio-pharmaceutically derived products although extreme sterilisation temperatures are required for thermal deactivation.

Drug dosage forms are made unstable by factors such as: temperature fluctuation, high water activity ('moistness'; A_w) and microbial intervention. Preservation both chemical and microbiological is usually achieved by combining individual techniques such as lowering water activity, and heat treatment. The quality of the drug product is defined by changes of the following types:

- *Physical* – transitions in drug polymorphs and iso-forms

- *Chemical* – evolution of impurities following catalysis and degradation

- *Microbiological* – increase in the microbial bio-burden.

The risk is removed by treating most products at temperatures greater than 80°C; specialised sterile products require treatment equivalent to five minutes at 121°C (250°F). Harsh processing environments are kept to a minimum at the same time as ensuring produce quality and safety. However, a combination of applied UV irradiation, ionising radiation, steam or ethylene oxide sterilisation, aseptic filling and preservatives can reduce a risk of spoilage and pathogen growth by using a 'hurdle effect' of inhibitory procedures.

System suitability indices

These really aim to give an indication of suitability and how much financial investment a method of scrutiny will require if done well. The specification of limits (see also analytical validation, Section 8.2) is applied equally across manufacture and laboratory testing. The five most essential and thus discriminatory indicators of a method's suitability are defined by the following:

- *Accuracy* – trueness to the real value (μ, x_{bar}) and resolving power; real accuracy is expensive to achieve (see Table 4.4)

- *Precision* – variability of findings (σ, σ^2, s, s^2, COV, SEM, etc.), important for setting the margins for acceptance and failure

- *Specificity* – the degree of discrimination of analyte A over similar analyte B

- *Ruggedness* – the extent of an ability to change experimental procedure without consequences

- *Robustness tests* – extent of method manipulation without loss of accuracy [Dejaegher and Heyden (2006)].

    Other valuable attributes of a method include the extent and range of linearity between measured attribute and quantity present, the selectivity and lack of influence of matrix components in a mixture on the result, the lowest concentration that can reasonably be determined without interference (sensitivity) and the limit of detection (LoD). The LoD represents approximately one tenth of the concentration to be measured and is important in trace analysis. It is important in all these cases to have defined the intended use of a procedure by reference to a spread of system suitability tests.

4.2.2 Modes of real practice

Chemical analysis and use of a suitable method also requires the use of suitable raw materials and standards. Use of certified reference material (CRM) and standard reference material (SRM) together with a certificate of analysis (CoA) that accompany materials tested by a contract laboratory are often used to reduce costs associated with QC testing in smaller organisations. As with all forms of assessment and examination reference to one of Deming's quotations is appropriate, 'cheaper is not always better *but* better is always cheaper' (see

Table 4.4   Representation of the qualities needed for accuracy and precision

Evaluation status/result quality	Accuracy	Precision	Data 'clumping' and similarity	True/ representative determined value	Comment
Very Poor	Low	Low	No	No	Systematic error (bias)
Poor	Low	High	Yes	No	Systematic error (bias)
Poor	High	Low	No	Yes	Random error?
Good	High	High	Yes	Yes	Error free

Table 4.4). This essentially refers to lower induced costs in the long-term based on consumers' and producers' errors (see later Section 6.1) during production and the product wastage associated with such misdiagnoses. The correct evaluation of raw materials, routine in-process samples, finished goods and stability samples relies on establishing benchmark standards for acceptance and failure that require pure chemical standards. Of course purity comes at a cost and so a manufacturer is required to balance the desired (ideal) test number, frequency and depth of scrutiny versus time, manpower and site capability.

Challenge testing

Manufacture-based challenge testing is an essential part of fundamental validation and constitutes the part of a performance qualification where the more extreme limits of use are tested. Most medicinal products are not distributed immediately following manufacture and so stored finished pharmaceutical products also require some degree of storage-based challenge testing to ensure the product remains as if it had just been manufactured.

To ensure the 'customer-suitability' of a product a number of worst-case-scenario conditions are usually established. These might include:

- Temperature changes ($\pm$20 per cent) for proteins (aggregation) and product containing sugars (amino-carbonyl reaction) such as lactose to ensure the product remains intact.

- Variations in lighting for retinoids, penicillins, B-vitamins, unsaturated lipids and polymers.

- Variations in humidity for hard gelatin/alginate capsules, pellets (non-pareil sugar beads). As high water activities tend to be associated with an increased rate of degradation over the solid state.

- Augmentation in oxygen saturation for unsaturated lipids and polyphenols, amino acids (biopharmaceuticals are particularly sensitive to both pH and oxygen, and stable over limited temperature ranges).

- Exposure to more extreme pHs for labile drugs e.g. aspirin, indomethacin, proteins.

- Exposure to air (microbiology) to give an indication of fouling.

- Exposure to mechanical actions that might allow denaturation of enzymes and peptides.

- The length of storage and abuse of the product shelf life.

- Extreme mechanical tests on packaging durability and perishing conditions that might be relevant to some goods e.g. PVC/PVdC blister packs.

In most cases the assessment of suitability should be made and tested on both product and packaging because this is the form of the medicine that the customer will receive.

Stability studies

These are undertaken for a number of reasons but the three most significant are to confirm that the product matches agreed standards of:

- Compliance and therefore an indication of process suitability

- Non-compliance and therefore an indication of product quality

- Specification to customer requirements.

In a standardised form, stability studies (Table 4.5) are undertaken on designated stability samples (not for distribution, as part of validation data accrual), and routine production pack samples. Production samples act as a way of testing what the customer receives. The standard testing regimes for products fit into three basic categories of 25°C (temperate climate), 35°C (sub-tropical climate) and 45°C (tropical climate) at higher relative humidity and under conditions of ambient artificial light. As a routine stability samples are assessed at the point of completion of manufacture and then periodically at three months, six months, one and two (and for complete data, three) years.

Table 4.5 Standardised stability test conditions for pharmaceuticals [MHRA (2002), 3–177]

Climate zone	Conditions	Temperature	Humidity
I	Temperate	21°C	45 %RH
II	Mediterranean/sub-tropic	25°C	60 %RH
III	Hot, dry	30°C	35 %RH
IV	Hot, humid	30°C	70 %RH
EXTREME		40°C	75 %RH

Accelerated light testing (1 million lux UV) is performed on samples that might be exposed to the more extreme forms of direct sunlight. Additional pressure testing might be relevant to those products shipped by air-freight. All samples are stored in 'drawers' by batch number in a stability room with limited access, which has temperature and humidity control and time records; these illustrate fluctuations to the established regime of storage. Some products may also require cooler temperatures [MHRA (2002); Ahuja and Scypinski (2001); FDA (1996) Freeman *et al.* (2003); Johnson (2003); Walsh and Murphy (1999); Doblhoff-Dier and Bliem (1999)] representing the likelihood that the sample (e.g. vaccines) may be stored under refrigerated conditions. Biopharmaceutical stability is often influenced by the process of lyophilisation, and humidity, oxidation and microbiological intervention to a greater extent than many conventional drugs [Walsh and Murphy (1999); FDA (1996); Sarker (2004)].

Product stability is undertaken for assurance of quality:

'. . . the purpose of stability testing is . . . EVIDENCE . . . on how the QUALITY of a . . . drug product [including API] varies with time under the INFLUENCE of environment . . . temperature, humidity, and light and enables RECOMMENDATION of storage conditions, re-test . . . and shelf lives . . .'

ICH, 1996

[Hora and Chen (1999); FDA (1996)]

Thus QA considers that safety and efficacy of the drug form are paramount. However, commercial interest, lost revenue, efficiency and ethical considerations are also important. The incorporation of stability studies into the testing of manufactured medicines also describes the overall supply chain in-process 'secured' points [Hora and Chen (1999)], particularly given the delayed release of medicine from manufacturer to outlet. The process analytics used to follow the drug product should be reliable and applicable across dosage forms in order to accrue validation data. However their real value lies in provision of a profile of product degradation, and *markers of product quality*.

Stability assessment according to [Robertson (1993)] is based on the reaction rate of drug catalysis and therefore its shelf-life degradation:

reactant ⇒ rate constants for the forward and back reactions ⇒ product.

Shelf life (quality, Q) is determined from chemical assay and sensory evaluation

$$\frac{\mathrm{d}A}{\mathrm{d}t} = kQn \qquad (4.5)$$

where A, t, k and n represent the amount of a marker, time, a constant, and the quantity of defectives, respectively. The formula provides us with a notion of the subtle point at which degradation is unacceptable by indicating a point at which a 'just noticeable difference' can be observed. Chemists and stability scientists are used to a definition called temperature quotient (Q_{10}). This is described as the rate of increase in a highlighted quality factor as a result of a ten-degree temperature increase over the base level; and in most cases significant increases are observed. Parameters such as this can be used to describe the quality loss in drug products.

Process control and description

Processes are controlled via batch documentation and batch records that chronicle standard and non-standard unit operations. Documentation following validation should detail all specific discrete steps within a given process. During any process appropriate process control points should be identified on the batch documentation. This facilitates a degree of systematic control over the quality of the end product.

5

Good manufacturing practices

The background to current good manufacturing practice (cGMP) and modes of operation has been the cost of 'carelessness' and a number of significant landmarks and benchmarks in acceptable quality. GMP has undergone an evolution [Hoyle (2006); Poe (2003)] from minimally tested product to one of product produced in the most tightly controlled environment possible. A synopsis of global control of quality in medicine manufacture is given below. This is particularly pertinent to the rigours of testing during clinical trials [Muller et al. (1996); Melethil (2006); Webster et al. (2005); Närhi and Nördstrom (2005); Crowley and FitzGerald (2006)]; after all the foundation of good manufacture is a drug substance which is fit for purpose.

Landmark stages in the evolution of cGMP have been:

- 1938 Federal Food, Drug and Cosmetic (FDC) Act (from FDA, 1906)

- 1962 FDA – Kefauver-Harris Amendments (thalidomide event)

- UK Medicines Act 1968, Misuse of Drugs Act 1971

- 1971 FDC revision

- 1978 FDC revision by FDA

- 1981 US, Environmental Protection Agency and WHO – provide essential stipulations on industrial pollution

- 1990s FDC revision/ICH established

- 1997 US/FDA statutes for essential provisions in drug manufacture (personnel, building, records, production, process-control)

- 2002 US/FDA statutes for work undertaken by contractors (including contract research, relevant to good clinical practice)

Quality Systems and Controls for Pharmaceuticals   D K Sarker
© 2008 John Wiley & Sons, Ltd

- 2003 uniformity in contract production in global market (Common Technical Document, mandatory EU/Japan, recommended US)

- 2010? ICH, international harmonisation of drug standards (at present it looks unlikely that the ICH will adhere to this timeframe)?

The cGMP requirements needed as a pre-requisite [Advanstar Communications (2003)] in modern day manufacture include the following essential functions:

- Conformance to MHRA, FDA guidelines

- Suitably qualified personnel

- Adequate premises (possibly making use of intelligent design), HVAC, HEPA filters various grades and a 99.99 per cent screening of 0.3 micron particles [Freeman *et al.* (2003)] for terminally sterilised medicines

- Suitable equipment (manufacturing and process control) and site services

- Correct materials including containers and labels

- QP approved procedures and instructions, such as SOPs used in all cases

- Suitable storage and transport of materials and end-product

- Batch recall procedures in place to prevent risk to the consumer

- Complaints regarding product failings examined and acted upon to improve the process

- Defects and deviations investigated (and recorded) and the causes of non-conformance identified with appropriate remedial steps taken

- Detailed records must be kept (this relates to electronic and hard-copy versions) to permit a detailed auditing of procedures. Product conformity and self inspection (auditing) is a requirement of cGMP [ISO (2000)] that relates to ISO9001:2000 (checking for systems compliance and variation).

It is worth remembering another of Deming's famous visionary quotes at this juncture that does well at embodying the primary goal of GMP, '*the customer* [patient or recipient] *is king*' and as such they define the appropriate degree of quality that is acceptable in a manufacturing process.

GMP considerations and the 'rigour of the control' on the process are to some extent influenced by the perceived risk to the product. The default position is to allow the minimum of risk and this does not always mean harsh chemical or physical treatment but can involve using multiple technological approaches, such as hurdle methods. There are a number of concerns for a range of medicinal products that are indicated in Tables 5.1 and 5.2.

Table 5.1 Pharmaceutical products and GMP considerations

Product/form	Examples	Risks*
Solid dosage – 70 %	Tablet, capsules, capsules containing waxed-sugar pellets (patches, etc.)	Raw materials (low water activity, A_w not maintained)
Liquid dosage – 25 %	Emulsion, suspension, etc.	Liquid-based microbial growth
Gaseous dosage	Aerosol, aphron?, medical gases	Liquid-based microbial growth

*–also purity and content – P (purity) C (consistency) Q (quality) [Sarker (2004)].

Table 5.2 Process concerns for three medicinal products

Product – Solid	Product – Dispersion	Product – Colloid
Tablet	Emulsion – cream, ointment	Nanoparticle suspension – chemotherapeutic
Preservation – low A_w	Preservation – limited water content, preservatives 'Hurdle mechanism'	Preservation – sterilisation, preservatives 'Hurdle mechanism'
Issues: Raw material (RM) quality Hygienic preparation	Issues: RM quality Hygienic preparation 'Pasteurisation', heating	Issues: RM quality Hygienic preparation Terminal sterilisation
Degradation and uniformity = conformity, efficacy	Degradation and uniformity = conformity, efficacy	Degradation, consistency, efficacy
Status: non-sterile Low risk	Status: non-sterile/sterile Low risk	Status: sterile High risk

GMP is essential in all cases from initial drug trials to commercial launch, however some preparations are easier to assemble, use materials that come from non-ideal sources (that can complicate matters further) and are easier to check. Some products are also easier to physically sample and thus obtain a picture of compositional inconsistency and variation. In order to obtain the best product a manufacturer *needs* a system (TMS) in place to ensure regular formulation, processing and composition (Table 5.3).

Without regulation of a manufacturing process the consequences resemble 'chaos' that might escape notice in the first instance but at some later point will invalidate the safety of the product. This means someone along the drug development and application chain gets injured. Safety of the patient is

Table 5.3   Key issues to be addressed as part of quality circle initiatives and GMP

Standards	Decisions	Improvements
Limits	Individualistic	Learn from mistakes
Materials	Altruistic	Try a new approach
Equipment	Informed	Additional improvement:
Facility	Group	1. Investigate/validate 'process' to nth degree
Methodologies	Responsive	
Team		2. Better communication and technology transfer
		3. Do not 'economise' on investigation and validation expenses

the most important consideration that drives the push for better quality and negates 'reasonable' expenditure on manufacturing. Pharmaceutical production profiles should be simple and *linear* in moving through the production unit operations. This is referred to as process linearity and best illustrated in Figure 5.1.

Some of the data from Figure 5.1 is taken from Sharp (2000) whilst other information is taken from personal experience of the best industrial practices. Space is essential to prevent accidents and transfer mistakes, including cross-contamination. Zonation is also essential to the optimal working of a factory or site of manufacturing and prevents cross-contamination events.

5.1 Manufacture of standard products

A 'standard product' in this sense is one where the unit operations and risk assessment of the end product suggest simple equipment and ambient conditions may be suitable. This does not, however, suggest that the product is manufactured in a casual or haphazard manner and that technologies involved in the manufacture are old-fashioned and over-simplistic. On the contrary it still means that a product is made according to a highly regimented and regulated procedure.

At this point it is convenient to divide drug products up into two generic categories and these are:

• Non-sterile medicines

• Sterile medicines, such as parenterals.

Non-sterile formulations carry with them an inherently lower risk which is associated with their point of application and the dry or low water activity of the dosage form (Table 5.2). As such they require only qualified person recommendation and sign-off on EACH batch (and compliance to PCQ issues). Sterile

(a) LINEAR						
Goods receipt (1)	Materials (2)	Dispensary (4)	Bulk products (6)	Packing (7, 3b)	Finished product (8)	Dispatch (9)
	Packaging (3a)	Bulk manufacturing (5)		[External link 3a and 3b]		

(b) U-FLOW				
Goods receipt (1)			Dispatch (9)	
Materials (2)	Packaging (3a)			Finished product (8)
Dispensary (4)				
Bulk manufacturing (5)		Bulk products (6)	Packing (7, 3b)	

(c) NON-LINEAR				
Goods receipt (1)	Dispensary (4)	Packaging (3a)	Materials (2)	It is clear in this chaotic set-up that there is an increased risk of cross- contamination, incorrect labelling and an unsatisfactorily high flow of personnel in between areas of varying risk. It is also worth noting that the environment is cramped.
Bulk manufacturing (5)		Finished product (8)	Packing (7, 3b)	
Bulk products (6)		Dispatch (9)		

Figure 5.1 Example of pharmaceutical manufacturing unit operations demonstrating (a) linearity, (b) U-flow – space-constrained form of linearity and (c) non-linear formats. Material moves as part of the assembly of product from a low number area (1) ultimately to a higher number area (9) in a sequential manner. The cells of the matrix represent individual unit operations and allocated space

medicines on the other hand, such as parenterals, are subject to post-process contamination and therefore often require 100 per cent testing (Table 5.1). This is routinely undertaken by optical or 'magic eye' sensors on dosage forms such as ampoular medicines; these look for particulates, colour changes or opacity as a means of assessing non-conformity. Destructive testing of the dosage form is obviously not a viable option.

Sterile medicines often include biotechnology derivatives where the consistency and potency of bio-preparations (which needs validation and constant monitoring) is often highly variable but may also be associated with issues of *purity:* incorporation of impurities and particulates. Routine testing of the drug often makes use of thermal analysis (calorimetry; DSC) and *analytical (microbiological) validation.* Where the risk of cross-contamination with hazardous pharmaceuticals is possible manufacturing suites usually only undertake one product per production campaign. In some cases such as with antibiotics the manufacture might involve facility dedication in order to cut down on unintentional adulteration. Sterile manufacture tends to be more rigorous in terms of equipment and specialised clean rooms. These specialised conditions and the nature of the drug itself often require additional and 'top-up' staff training and a stronger reliance on the QP to sign-off (accept) or reject batches of medicine. Tablets-capsules-pellets represent approximately 70 per cent of all medicines, with topicals at 20 per cent and parenterals at 5 per cent (see Table 5.1). As such there are different risks associated with the extent of their use. This is an over simplification to some extent because although represented by a mere 5 per cent of formulations certain high risk therapies (chemotherapy, vaccines, etc.) make use of a disproportionate number of parenteral routes of administration and so parenteral products can be considered to represent the higher risk category.

5.1.1 Solid dosages: tablets, capsules, pellets and soft-gelatin capsules

Solid drug forms are used for gastric, enteric, sub-lingual and buccal applications (Figure 5.2). Pills (round tablets) and lozenge-shaped tablets represent the vast majority, >65 per cent, of applications of these medicines. Finished goods such as non-pareil sugar pellets in capsules, tablets and capsules must comply with specifications [Colombo *et al.* (2000); Smith (1999); Breitenbach (2002); Nazzal and Khan (2006)]. The general rule for product suitability is that the dosage form should meet the following criteria:

- QUALITY (microbiological status, product form)
- QUANTITY (potency)
- PURITY (consistency)
- CONTAMINANT-FREE.

Compressed tablet manufacture is based on granulation of active drug with formulation and bulking aid excipients. These excipients: solids, polymers,

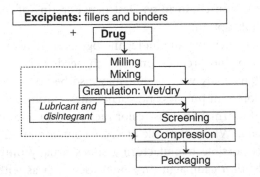

Figure 5.2 Highly generalised schematic of generic tablet manufacture. Granulation can involve a wet or dry process and this depends on the characteristics of the drug and excipients. Capsules would miss out the compression stage and milled granulate would pass to storage then be dosed directly into the capsule shell

moisture, water-soluble sugars, influence the manner in which the solid drug particle is coated or bonded to neighbouring particles. This granule formation is achieved via wet or dry granulation, drying, milling and screening (12–20 mesh sieve for tablet granulates equal to 2 mm to 900 microns) then compression [Forster *et al.* (2002); Sonnergaard (2006)]. Granulation is a key step as this determines particle size, shape and ultimately dosage form hardness. These factors influence the compressibility of the granulate powder, which becomes more significant when multiple compression stages are involved. Modern processing makes use of fluidised bed drying and oven-based methods. Non-pareil waxed pellets for sustained release capsules are coated with poorly soluble waxes to give a particular release profile; this is done in a combined blender/drier. Other uses for milled granulate may be for sacheted pharmaceutical powders or metered inhalation devices, in this case small size (1–6 microns) and shape of the particle are crucial to optimal product performance in the lungs. Tablet coating is now customarily undertaken in an industrial scale sprayed tumble blender drier, rather than a coating pan. Some tablets require the additional steps of glaze addition, polishing or enteric coating [Ansel *et al.* (1999)].

Processing conditions such as granulation, drying and compression times, and temperature profiles: long–high, short–high, short–low and long–low, have huge influences on the finished goods. The processing behind manufacture of the raw material active, such as fluidised bed drying and lyophilisation, also has a significant impact on end product variability and hence quality.

Solid dosage form ingredients include the active pharmaceutical ingredient (API), fillers, binders, dissolution aids, solubilisers, lubricants, disintegrants, glidants, coatings; also making use of granulation aids (IMS, methylene dichloride, ethanol, water-for-injection). Common ingredients might include:

- Fillers – calcium carbonates, phosphates and sulphates, starch, milled celluloses (MCC), lactose

- Binders – polyvinyl pyrrolidone (PVP), sodium carboxymethylcellulose (CMC), gelatin, polyvinyl alcohol, methacrylates

- Controlled and sustained release agents – poloxamers, polyethylene glycol (PEG), hydroxypropyl and hydroxyethyl methyl celluloses (HPMC; HEMC)

- Lubricants – Zn/Mg/Ca stearates, talc, PEG, sodium lauryl sulphate (SDS), vegetable and mineral oils

- Disintegrants – starch, guar gum, kaolin, alginates, bentonite, silicates such as Veegum

- Glidants – aerosil, cab-o-sil fumed silicas

- Coatings and colourings – polymers and surfactant solutions, sugar/wax; synthetic and natural dyes (FD&C colours), lakes (alumina) and pigments (ferric oxide grades, titanium dioxide)

- Additives – humectants (sugar alcohols such as sorbitol), buffer salts, de-humectants, antioxidants.

Structuring within the granule and therefore the tablet is important as this can influence the mode of drug release by either erosion, swelling of a polymer matrix [Colombo *et al.* (2000)] or dissolution and perhaps even a combination of all three processes. Sustained release, which allows gradual release of the drug, or controlled release, which permits release in a specific location or environment, can be and are used frequently for the delivery of drugs using oral formulations. Recently interest has been sparked by the use of 'glass solutions' formed by melt extrusion (solid dispersions) for poorly soluble drugs such as Griseofulvin in PEG (Grispeg-Sandoz), nabilone in PVP (Cesamet-Lilly) and their use to avoid the complex granulation process [Forster *et al.* (2002)].

With solid dosage forms the key to controlling intrinsic variation and thus qualities are:

- Tightly controlled granulate drying times and temperatures

- Tightly controlled humidity (relative humidity, product A_w) and solvent or co-solvent content and use

- Tightly controlled particle size and shape (distribution), type of granule and cross-links formed as a result of wet granulation and screening. Granulation time, milling time/screen size are clearly defined

- Particle compressibility, fracture strength, tensile strength and elasticity

- Interplay of excipients and component chemical compatibility

- Form and ratio of soluble to insoluble material

- Tablet compression ratio and the type and speed of compression are controlled and reviewed periodically during the production run

- The thickness, solubility and hardness of coatings is evaluated

- Chemical inconsistencies are evaluated

- The raw material and end-product storage time/temperature/profile are controlled.

The dimensions of solid dosage forms (tablets) are determined by punches and dies. Appropriate protection from wear and tear of tungsten carbide punches with raised impressions is vital given their expected lifetime (for example > 30 000 basic tablets per hour over say 100 days use is expected to create some wear), precision engineering and use. Disintegration, content of uniformity (see analytical validation) and weight (USP weight variation), thickness, hardness/friability, compression tests and USP (US pharmacopeial) dissolution testing are essential standard QC tests. Other tests would include freedom from pathogenic micro-organisms and quantification of chemical impurities. Visual inspection and compliance with product licence including for monographing, break bars and colour before and after coating are also needed. Since the tablets and capsules are often presented in blister packs, blister pack compromise tests would also be appropriate and might include blister burst tests.

Tabletting involves the manufacture of simple compressed tablets (cores), multiple compression forms, sugar-coated tablets, film-coated tablets and enteric-coated tablets. Additionally tablets can also include sugar-based solid dispersions of drugs usually based on sucrose or di/trisaccharide sugars [Forster *et al.* (2002)], although this represents a technology that is not universally used or available yet. Specialist use tablets include buccal, sublingual, effervescent, chewable and immediate release forms that have formulations appropriate to their use and should be checked for suitability to specific application via appropriate SOPs. These may also be relevant to buccal delivery in chewable form. Some examples of tableted medicines include Acyclovir (antiviral, GSK) larger dosages at 800 mg, Lovastatin (anti-hypercholesterolemic, Merck) 40 mg, and Levothyroxine (Na – hormone, Knoll) 0.025–0.3 mg dosage strengths. Obviously there are concerns of content of uniformity and accurate dosing in cases where the dosage strength is low.

Hard shell capsules are influenced by humidity/temperature and can vary in size, diameter, colour coding, the ease of capsule cleaning and polishing and ranging from sizes 000, 00, 0 to 5 [Ansel *et al.* (1999)]. They are produced in an automatic capsule filler (>165 000/h) that works well if the alignment of granulate and shell are matched. Examples of the range of active contents for capsules include Amoxicillin (antibacterial, Wyeth-Ayerst) at 500 mg, fluazepam HCl (hypnotic, Roche) 15 mg capsules and indomethacin (anti-inflammatory, Merck) at 50 mg.

Soft 'gelatin' capsules require careful alignment of feed and sealing rates of the feeder rolls of capsule material and the injection of active before sealing by the die. Some products formulated in soft capsule form include: Cyclosporine (antimicrobial in microemulsion format, Novartis), Ranitidine HCl (emulsion format, GSK) and Digoxin cardiotonic (GSK) in polymer dispersion. The use of soft gelatine capsules is not nearly as commonplace as hard gelatine capsules.

5.1.2 Transdermal delivery and dressings

These products exist in a number of forms ranging from hydrogels to emulsions and organogels [Florence and Attwood (1998)]. Transdermally used emulsions (transdermals) are formulated to have a moderately high surface zeta-potential (+40 mV) and are therefore not susceptible to coagulation. This surface charge is achieved by using blends of cationic lipids and surfactants (emulsifiers). In such a case the formulation would typically include:

- Fatty acids and phospholipid mix (primary 'emulsifier')

- Poloxamine 188 (neutral polymeric emulsifier)

- Cationic stearylamine ($pK_a \approx 11$)

- α-tocopherol (preservative/anti-oxidant)

- *Drug*: such as nicotine for use in nicotine patches.

Human skin is composed of the *stratum corneum* (SC; 20 per cent lipids, 40 per cent β-keratin and about 40 per cent water), *dermis* (75 per cent collagen), sub-cutaneous fat and vasculature and superficial muscle according to a 'brick wall' hypothesis where the cells represent the bricks [Moghimi (1996)]. The rate and extent of drug delivery across the SC is dependent on [Florence and Attwood (1998)]:

- *In situ* chemical modifications that take place within the skin

- The absorption profile and ease of take-up into the skin layers

- The vehicle type that might include an oil base, w/o or o/w emulsion type

- Chemical properties of the drug product itself in terms of hydration, temperature, pH, drug concentration and penetrant

- Intrinsic skin variations.

Occlusive films prevent water loss that might include corticosteroid preparations and thus aid penetration between the lipid portion and the proteinaceous portion of the skin. The vehicle has a very powerful contribution to drug passage because it can cause emulsion inversion (altering penetration), volatilisation and precipitation and loss of efficacy. Consequently, appropriate delivery of the drug usually appears in the form of ointments (waxes), absorption bases (organo- and hydro-gels), such as suppositories and aqueous creams (semi-solids: bases and emulsions).

The advantages of transdermal and patch technologies include:

- Removal of inconsistencies of gastric absorption (time, pH, matrix)

- The form can allow systemic circulation without direct passage into portal system

- The dosage can be administered as a contrast or continuous supply

- The method eliminates the 'pulsed' delivery of the drug

- Removal of the patch terminates absorption rapidly.

Disadvantages of transdermal and patch technologies include:

- Inconsistency; different drugs are absorbed more or less depending on skin and drug chemistry

- Nicotine or oestrogens are used routinely because the absorbed dose is low (flux $0.08 \, mgcm^{-2}s^{-1}$)

- The absorption flux, however, is susceptible to small variations in formulation (polymer blend) and skin pH changes.

Commercial examples of patches include Estraderm® (oestradiol in poly (isobutylene)) and Nicotinell® (nicotine in acrylate) both produced by Novartis. These and other forms of transdermal delivery have been and continue to be used with great success. Basic control tests would involve measurements of drug content degradation of the patch and customarily some assessment of the lifetime of the product.

Wound management makes use of 'dressings', which can take many forms. Although they are not (or not always) medicines they do often elicit a non-specific medical effect; however they are not subject to the same controls as medicines. They may be used alongside formally controlled medicines but are subject to an EU Medical Device Directive in Euro-zone countries that require a suitability 'CE' stamp an all products, verified in the UK by the MHRA. The directive lists requisite details and ranks fours classes of product, I (low risk), IIa (low–medium risk), IIb (medium risk, complex-natured product) and III (high risk, incorporates medicine). Dressings include: casting materials, bandages, tapes, adsorbents and wound management products (these are diverse and can take many specialised formats). In the UK testing would be part of those listed under Appendix XX of the BP (1993). These basically detail material identification and textural properties relevant to the product use such as tensile strength, and physical tests such as moisture content or absorbency, barrier properties and 'in-house' tests. Appropriate tests are undertaken under the supervision of the QP.

5.1.3 Non-sterile liquids and paediatric syrups

These are usually stabilised by high quantities of sucrose or other monosaccharides and by acidifiers and food-grade preservatives, such as ethanol and benzoates. Particle size is an important consideration (see also Table 5.6; Section 5.2.1) as this influences settling and sedimentation time and can thus influence dosage per any unit of dosage. The primary consideration would be the presence of pathogenic organisms but this is negated by application of 'hurdle technology' as in pasteurisation, high sugar content and preservative action which serve to considerably reduce the number of organisms that can tolerate the environment [Hanlon (2002); Hodges (2002)]. Basic control tests would include active content, particle size as this might influence sedimentation rates, refractive index measurements of sugar content, and freedom from pathogens and foreign matter. The likelihood of short shelf life following opening needs to be considered.

5.1.4 Topical (emulsion) products, including medicated shampoos

Topical products represent about a fifth of all administered medicines. These products are inherently unstable to temperature variation, pressure changes, long-term storage and microbial growth. Consistency varies with excipient purity and the nature and fluidity of the dispersed component and the dispersion medium. Sterility is not always required and this determines the effective shelf life. Sterility by thermal means is hard to achieve as this destabilises the dispersion and produces a different consistency on cooling. To circumvent this problem hot manufacture at 70–80°C is usually undertaken. However products should be pathogen free particularly as the product may be applied to broken skin.

Emulsions are used where the formulation includes:

- Poorly soluble compounds (drugs e.g. steroids; lyophobic biomedical agents e.g. imaging agents; flavours and colouring e.g. essential oils).

- The need for encapsulation and retention (see parenterals Section 5.2.1) of labile or, crucially, cytotoxic drugs.

- Specific delivery and purposes such as epicutaneous, rectal or vaginal delivery. This can be by suppositories or even by foam [Klotz and Schwab (2005)]. The foam technique has been used to deliver 5-aminosalicylate or the azo-pro-drug (sulfasalazine) for ulcerative colitis (Pentasa®, Salofalk®) and the corticosteroid budesonide, metabolised in the rectum. Emulsified products may also make use of 'aggregates' of which there are two basic types: associates and coacervate complexes.

Emulsions are variously described as coarse dispersions, regular emulsions, micron-sized emulsions, and solid lipid micro-capsules, depending on their size and components. In the strict sense topical products of this type are metastable and given enough time will split and revert back to their original unmixed form. In this manner they are different to micelles, microemulsions

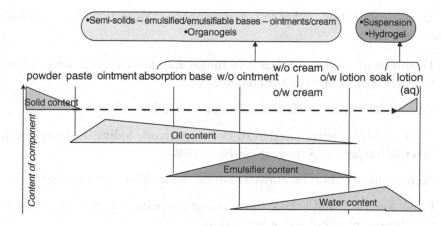

Figure 5.3 The pharmaceutical emulsions and dispersions family

Table 5.4 Emulsified drugs past and present [Klang *et al.* (1998), 31–65]

Nature of drug or chemical	Examples (therapeutic application)
Lipophiles-hydrophobes	Amphotericin B (systemic antifungal)
	Diazepam (sedative/hypnotic)
	PGE$_1$ (prostaglandin, vascular therapy)
	Dexamethasone palmitate (rheumatoid arthritis)
	Propofol (general anaesthesia)
	Tocopherols – vitamin E (nutrition)
Cytotoxic and harmful species that do not formulate in aqueous form	Penclomidine (chemotherapy)
	Rhizoxin (chemotherapy)
	Podophyllotoxin (chemotherapy)

and true colloidal dispersions, which are thermodynamically stable and optically transparent, requiring no mechanical agitation for them to spontaneously form.

Topical products (Figure 5.3) can also include foams, such as shampoos and bathing foams. Typically, these involve incorporation of drug (Table 5.4) or emulsified or polymer entrapped drug (and possibly pro-drug form). Altering droplet or oleosome surface rheology can be used to alter the release time [Guy *et al.* (1982); Washington (1990); Speiser (1998); Müller and Böhm (1998); Sarker (2005b); Floyd (1999); Riess and Krafft (1998)] and in this manner control product efficacy. Topical preparations are used for a range of medicines such as corticosteroids (betamethasone valerate, phosphate (Luxiq™, Connetics Corp.) approved by the FDA and clobetasol propionate for scalp dermatoses and psoriasis (betamethasone diproprionate); they can work better than conventional gels and solutions [Feldman *et al.* (2000)]. In some cases such as betamethasone they can be applied in foam form as Bettamousse (Celltech), which exists as an emulsion-foam hybrid. Other topical preparations can also include medicated shampoos for:

- Treatment of psoriasis, dermatitis – fungal and dandruff (e.g. ketoconazole (status being reviewed) and fluconazole)

- Treatment of seborrheic lesions (selenium sulphide)

- Treatment of bacterial infection (triclosan, povidone-iodine (*Ps. aeruginosa*), silver sulphadiazine (*S. aureus*) [Snelling (1980)]

- Bacterial management for burns (chlorhexidine, sulphur (use to be reviewed))

- Pediculosis (lice) and scabies; these contain the pesticides lindane/malathion, pyrethrums (plant extract) or permethrin.

Ingredients in dermatological shampoos can include 'salicylates' (Targum™ shampoo), selenium, sulphur and coal tar (Zetar™ shampoo) for the routine treatment of psoriasis [Shapiro and Maddin (1996); Snelling (1980)]. Further uses for emulsified products are also discussed in the section on parenteral medicines, which use smaller droplets than those used in the coarse dispersions for topical administration.

Examples of topical medicines used routinely include hydrocortisone cream (0.1 per cent w/w BP) and similar formulations. Here, the hydrophobicity of drug is critical, as this would influence encapsulation ratio and thus the potency per unit volume of the dosage form. For example, it is considered that the drug would not be suitably encapsulated in the dispersed oil phase if log P values for the moiety were <0.8 [Klang *et al.* (1998)]. One strategy might be to form a pro-drug chemical conjugate with fatty acid-cholesterol or otherwise derivatise the molecule to increase its lipophilicity; in the former case log P goes to a value close to 20.

Topical products include aqueous and non-aqueous (organogel) gels and their use is widespread. The uniform consistency and drug content of simpler forms such as these are points of consideration in maintaining quality. In general they do not form as diverse a group of drug delivery systems for topical administration as emulsions. Emulsion function is defined by the individual components indicated in Figure 5.4. In the majority of cases uniformity of droplet size, the polydispersity of droplets, particle drug content and the texture of the dispersion medium have a key role to play in product stability [Sarker (2005a); Martin (1993); Becher (2001)]. Modifications to storage temperature, possible temperature cycling and alterations to microbial growth (evolution of biosurfactants) and pH alterations could also be considered to play a vital role in maintaining or altering the product shelf life. Routine QC tests for the product should include rheological measurements, content of uniformity, particle sizing and pH measurement. The products (creams, lotions, absorption bases, etc.) make use of a wide range of

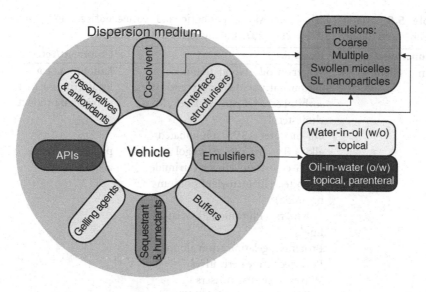

Figure 5.4 Key topical product ingredients and excipients

active ingredients that are used in both topical and often parenteral applications. Some of the ingredients are also used for cosmetics and are indicated in Table 5.5.

Dispersions using solid, liquid and low-boiling-point liquid phases (such as the Oxygent® parenteral system) or semi-solids [Klang *et al.* (1998)] are very often based on emulsions. Emulsions are used for topical, parenteral, patch, soft gelatin capsules and ocular systems. The surfactant may be considered to be the most important primary ingredient (other than the drug).

Droplet size maintenance is essential to product stability since polydisperse drug products of this type are prone to creaming, Ostwald ripening (dispropor-tionation of oil droplets) and coalescence. The size of particles in the dispersion can be crudely assessed by reference to a 1–10 per cent dispersion of the parent product. The particle size is revealed in terms of the colour or opacity observed.

Customary control and assessment of particle size is undertaken by light scat-tering (photon correlation spectroscopy) technologies such as the Mastersizer™ or Zetasizer™ from Malvern Instruments. Many commercial instruments exist which undertake predictions of size based on the scattering of light or other basic physical properties.

Products that make use of dispersed oil or more rarely dispersed water because of problems of biocompatibility (see the section on transdermal delivery and the composition of the skin) are commonplace. Coarse emulsions (emulsions, general emulsions) are usually 1–100 microns (μm) in diameter and they are characterised by being thermodynamically unstable (metastable) and susceptible to phase separation (unsightly splitting over long storage periods). They are however, very effective at the solubilisation of drugs and presentation in a

Table 5.5 Pharmaceutical emulsions, cosmetic and cosmeceuticals excipient use [Banker and Rhodes (1979), 263–357; Billany (2002), 334–359]

Functional agent	Example	Comment/reference
Emulsifier	Water-in-oil – woolfat/lanolin, anionic fatty acid salts, glyceryl esters, sorbitol esters, Spans, Arlacels, Myrj, cholesterol	Lanolin can cause local irritation as a parenteral product
	Oil-in-water – SDS, sulphonated alkyls, Tweens, cetomacrogol (PEG-cetostearyl ether), cetrimide (cetostearyltrimethylammonium bromide)	pH change using alkali soaps?
	Polymers – Pluronics, PEGylated lipids	
	Proteins – gelatin/serum albumin	
	Polysaccharide derivatives	
	Waxes – see structurisers	
	Solids – Veegum, solid lipid nanobeads (SLNs) (Lipoperls™), NanoCrystals™	
	Formulations – precirol ATO-5 (i.e. glyceryl derivative lipids, Pluronic F68 (poloxamine 188)	
Co-solvent	Liquid poly(ethylene glycol) (PEG), glycerol, propylene glycol, sorbitol	
Structurisers	Waxes – Carnuba wax, beeswax – cosmetics; higher fatty alcohols and derivatives – pharmaceuticals	Slow down creaming or can cause flocculation. Add textural properties to the dosage form. They can be lyophobic (more rarely used) or more customarily lyophilic (hydrophilic).
	Polymers – silicones, alginates, tragacanth, acacia, ethylcellulose, hydroxyethylcellulose, carboxyvinyl polymer (Carbopol)	
	Solids – Veegum, fumed silica (Cab-o-sil), silicones, bentonite, lamponite, montmorillonite, carbon black	
Gelation aiding agents	Ethanolamine, triethanolamine	
Vehicle	Cosmetics – Squalene oil, as pharmaceuticals	Petrolatum comes in light, heavy soft and hard grades
	Pharmaceutical – liquid petrolatum (paraffin), mineral oil, oleyl alcohols, silicone oil, soybean oil, arachis/sesame/cottonseed/castor oils, hydrogenated/sulphated oils	

Table 5.5 *(continued)*

Functional agent	Example	Comment/reference
Minor additives	Sequestrants – citric acid, EDTA Buffers – glycine, acetic phosphoric acid/salts Preservatives – butyl/propyl/methyl parabens, benzyl alcohol Antioxidants – BHT/BHA/propyl gallate, tocopherol, ascorbyl palmitate Humectants – xylitol, glycerol (glycerine), sorbitol	Used to maintain the product integrity

Table 5.6 Approximate sizes for pharmaceutical emulsions or solid aqueous suspensions [University of Florida (2006)]

Droplet diameter	Visual appearance, for a diluted suspension
$>1\,\mu m$	Milky white
$0.1–1\,\mu m$	Blue-white
$0.05–0.1\,\mu m$	Grey, semi-transparent
$<0.05\,\mu m$	Transparent (microemulsion domain)

uniform manner as long as the droplet size is small. As a general rule droplet sizes of less than five microns (5000 nm) give products with a longer shelf life. Droplet size, form and coverage are also important in terms of the drug release time, since there are multiple kinetics of release [Washington (1998)] for different size distributions and effective permeabilities of the emulsifier used to coat the surface. In many cases the exact location of the drug within the droplet of adsorbed surface layer is unknown and this can have a great bearing on the scope for catalysis and drug hydrolysis. The emulsifier coating (often sorbitan esters and polysorbates) is important as the palisade layer of surfactant (emulsifier) can influence the extent of coalescence and flocculation [Sarker (2005a)] as indicated in Figure 5.5.

Emulsions are inherently unstable but we can talk about their stability in terms of how long they can exist in a suitable dispersed form. Thus, the general stability of a topical is maintained by inclusion of other ingredients (excipients) in the 'stock' emulsion (Table 5.5; Figure 5.4). These shelf-life extenders and preservatives include salts and tonicity agents ($\sim$0.5 per cent). They are included for two reasons: primarily for the purposes of biocompatibility but also to act as particle stabilisers and charge screening aids that might prevent droplet coagulation. Tonicity agents commonly include dextrose, common salts (KCl and NaCl), buffers ($NaHCO_3$) and isotonicity or gelation aids ($CaCl_2$, $MgCl_2$) [Klang *et al.* (1998)]. In addition to simple emulsions two other types of emulsion exist: (1) multiple emulsions and (2) microemulsions. Microemulsions are colloidal particles and will be dealt with elsewhere (Section 5.2.1).

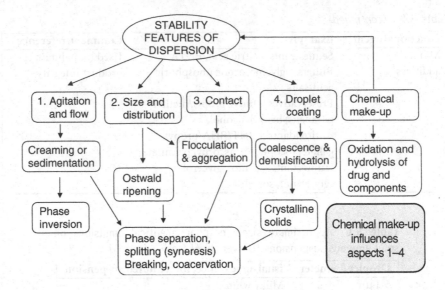

Figure 5.5   Emulsion and dispersion stability

However, multiple emulsions are used even if rather limited in number for encapsulation purposes. They can consist of water dispersed within oil, which is dispersed within water (W/O/W) or other formats such as O/W/O. There fabrication is based on using emulsifiers with differing shapes and hydrophobicities. As a general rule they are not commonly used for pharmaceutical drug delivery but essentially involve mixing two types of dispersions to produce a new format.

The applications of topical products, of which emulsions (or solid coarse dispersions) form the majority, are widespread and demonstrated in Table 5.7. Due to their relatively simple form of manufacture and 'universal' applicability emulsified liquid or solid lipid particles find use in anaesthesia, emolliency,

Table 5.7   Emulsion functioning [Klang *et al.* (1998), 31–65; Huang *et al.* (2005), S26–S38]

Delivery form	Bioavailability improved by emulsification	Controlled or prolonged effect observed
Ocular	Yes	Yes*
Transdermal	'Emulsion' usual form	Yes by residence in the upper layers of the skin
Oral	Yes	Yes
Topical	Increased skin permeation	Yes*

* – depending on formulation characteristics.

standard forms of drug delivery, hydrophobe encapsulation and controlled release applications.

5.2 Manufacture of materials requiring specialised production facilities

In addition to the rigorous GMP conditions required for non-specialised products, additional safety measures and precautions are needed; included in this category are drugs administered as aerosols, suspensions and solid powders such as those based on water-soluble particles of lactose (solid dose inhalers) for application via nasal and pulmonary routes. This might necessitate specialised delivery forms or sterile products. For dispensing devices [Ishizuka *et al.* (1995)] suitable quality control needs to be undertaken on device functioning and dose delivery. Blow-fill-seal MDPE, polycarbonate technologies and packaging materials for pharmaceutical use, that range from 0.1 ml to 1 litre container sizes (for aseptic liquids), using plastics and pharmaceutical grade borosilicate glasses, are now commonplace.

5.2.1 Parenterals

These products account for roughly less than 10 per cent of marketed pharmaceutical products but their use is often highly specialised and often requires considerably more caution because of the lack of 'barrier', and direct form of administration. Parenteral medicines are used universally in intra-venous forms for chemotherapy. The types of products can include dispersions of proteins, micellar and solid lipid aggregates and liposomes. The standard forms of systemic delivery for this class of product include:

- The skin – intra-venous (*iv*), intra-muscular (*im*) and sub-cutaneous (*sc*) depot

- Miscellaneous – intra-osseous (*ios*), intra-synovial (*isy*), intra-spinal (*is*), intra-cardiac (*ic*), intraocular (*io*), intra-respiratory (*ir*); and these are just a few examples of applications.

Parenteral products have strict ISO standards for any processes of manufacture and particularly for cleanliness of the production suite [MHRA (2002)]. The prime considerations for preparation of such products include the suitability of premises and equipment (to safeguard minimum levels of contamination), sanitation and waste disposal, and aspects of prolongation or rapid processing and handling. The cleanliness of the environment is usually defined by the quantity of airborne 'particulates' (these can represent dust, drug, or microbes).

The more stringent standards apply for 'clean-rooms or aseptic manufacture' because the medicinal products are either delivered ocularly or systemically. The size of particles in the range $0.5-5\,\mu m$, and in a cubic metre of air is counted to give an indication of cleanliness. Various standards based on the number of particles are indicated and assigned classes:

- Class 100 (100 particles of the size indicated above are present; in other classifications grade A/B are equivalent to class 100, they are distinguished by $5\,\mu m$ particle air loads during operation rather than at rest)
- Class 1000
- Class 10 000
- Class 100 000.

According to US Federal Standard 209E the cleanliness of a production environment can also be defined by an equivalent alphabetised labelling system. Here, grade (class) A (for high risk products), grade B (aseptic filling), grade C and D (lower risk products) are again used to define the 'freedom' from solid particle or microbial contaminants. Parenteral manufacture routinely uses the notion of segregation and controlled access zones. These are usually given labels such as 'cold, medium and hot risk' or more familiarly, low-moderate-high risk zones associated with the intrinsic risk of cross contamination of the product. Each zone is segregated and completely inaccessible to personnel that are not dedicated to working in one area; in addition material can only move linearly from one zone to another and is not permitted to pass in the other direction. This controlled process ensures that material lying in one area remains entirely in that area.

Microbial limits for cleanliness are similarly assigned grades A–D [Bloomfield and Baird (1996)] and these are based on pour-plate, settle, air and glove microbiological assays. The grade number relates to number of particles present, as grades A–D represent a scale from <1 to 200 cells/m^3 of air, respectively. Conventional clean room facilities use isolator technology to ensure reduced bio-burdens. One routine microbiological test for parenteral products would be for microbial pyrogenic compounds. These are evaluated by the limulus amoebacyte lysate (LAL) test that measures the clotting of a horseshoe crab blood test material. Parenteral products require the lowest microbial load for terminal sterilisation; only products with a load of better than grade A are suitable. The sterilisation may typically involve moist heat (autoclaving, retorting) or irradiation using cobalt-60. This may be a problem for particular types of parenterals e.g. microemulsion and emulsions such as Intralipid, since thermal treatment may destabilise the formulation, initiate chemical instability or drug polymorphism; consequently multiple methodologies of preservation such as aseptic filtering and chemical treatment may be considered as alternatives for producing high quality injectables.

Parenteral products are best designed to avoid precipitation of the drug at injection site [Klang *et al.* (1998); Washington (1998)]. This is important for drug products bearing cytotoxic drugs such as Taxol (paclitaxel) from Bristol-Myers Squibb, and as such they are often administered in encapsulated form (liposomes, emulsions, microemulsions, see Section 5.2.1) rather than suspensions of the drug alone. Carrying drugs in emulsions for parenteral use was initiated in the 1960s by the firm Kabi (Pharmacia) in Sweden, when they developed their dosage form Intralipid® for parenteral nutrition, with a tiny particle diameter of 70–400 nm that ensured long-term dispersal. The formulation made use of an amalgam of soybean oil, egg yolk phospholipid (food additive E322; lecithin), emulsifiers and glycerol. Ingredients such as these are susceptible to oxidative changes and should be carefully evaluated for their purity, consistency and quality. Parenteral therapeutics involving emulsions include further adaptations of the Intralipid formulation:

- Diprivan Emulsion® – using the anaesthetic proprofol (ICI/AstraZeneca), dating from the 1980s

- Vitalipid® – parenteral nutrition with lipophilic vitamins A, D, E, K (Kabi Nutrition)

- Limethason® – for the treatment of rheumatoid arthritis with dexamethasone palmitate (Green Cross Pharma)

- Lipo-NSAID® – which uses the chemical conjugate 'pro-drug' flurbipro-fen axetil (active flurbiprofen), a non-steroidal anti-inflammatory analgesic (Kaken Pharma).

Other adaptations based on 'stable' emulsions for parenteral administration include Diazepam (Diazemuls®) for sedation (Kabi) with derivatised lipids, a prostaglandin dispersion (Liple®) for vascular disorders (Green Cross), a perfluorooctylbromide (Oxygent®) blood substitute (Alliance Pharmaceutical Corporation) that makes use of block copolymer surfactant and a radio-imaging agent (Lipiodol®) that uses drug conjugates such as doxorubicin and epirubicin (Laboratoire Guerbet, France).

Long circulating parenteral emulsions require longer blood residence and avoidance of capture by mononuclear phagocyte system (MPS) organs (liver, spleen, bone marrow). Clearance can be a problem where delivery is to organs such as the brain. This can be facilitated by replacing egg lecithin in the formulation with block copolymer, polysorbate and poly(ethylene glycol) surfactants. Circulatory lifetime can also be prolonged by reducing particle size (<1 micron) and using charged surfactants to stabilise the droplet surface. The purity and consistency of excipients is exceptionally important in avoiding product inconsistency and the impact this might have on both quality and efficacy. In this manner the constancy and provision capability aligned with the integrity of the supplier become important considerations.

Solutions and suspensions

The major concerns with solutions are the buffering, API concentration and freedom from pathogens. These are available in a range of packaging from HDPE and MDPE copolymer plastics to plastic and glass ampoules of various sizes. All materials are sterilised at 121°C under pressure for sufficient time to destroy the presence of vegetative cells and spores of *C. botulinum* and indicator thermophiles, such as *B. stearothermophilus*. In this case thermal degradation of the API (active; drug) has to be carefully monitored by QC. Aseptic filling facilities permit use of cobalt-60 irradiation and the complementary methods of thermal and aseptic filling using 0.22 μm pore polymer membrane filtration. Solutions of radiopharmaceutical and sterile liquids are usually produced in a clean-room environment and may also use ethylene oxide to remove viable vegetative cells by fumigation of the clean room (see Sections 5.3 and 5.3.2, specifically Table 5.13 and Table 5.14). Examples of these types of products include magmas, irrigants, liniments and syrups [Ansel *et al.* (1999)]. Isotonic solutions for intravenous (*iv*) use are used with 'improvers' for biocompatibility purposes and typically these might include 0.9 per cent NaCl and 1 per cent dextrose incorporated into the solutions. Preservation is achieved by use of sequestrant (for example, EDTA) and is particularly of use where salts might be included in solutions for isotonic purposes.

Imaging suspensions may include ferro-magnetic fluids, ingredients such as radio-isotopes (technetium, indium) and magneto-pharmaceuticals. These products often make use of silica and magnetite for support of the active ingredient. These types of drug product are used in MRI (computed tomography) and X-ray applications. Other opaque materials used for imaging purposes include ingredients such as barium sulphate that might be incorporated into a barium enema dispersion (see emulsions within the topicals section, 5.1.4) or gastric meal. As with all coarse dispersions subject to both creaming or sedimentation suspension, particle size and the stability of aqueous buffered solutions needs to be carefully controlled in order to prevent dosing problems.

Colloidal mixtures and biomedical nanotechnology

Colloids (nanoparticles) and colloidal dispersions can span the particle size range from microns to angstroms, however most components are in the range of hundreds of nanometers ($\sim 10^{-7}$ m). The entities can range from structured to amorphous configurations and from mono- or multi-component crystalline bodies to association structures, and alternate between water miscible and water incompatible forms. There are a wide number of sources of information on this rapidly growing area of interest, which has begun to spawn a wide range of new patented drug dosage forms [Nielsen and Gohla (1998); Müller and Böhm (1998); Shaw (1992); Sinko (2006); Goodwin (2004); Hiemenz and Rajagopalan (1997); Valtcheva-Sarker *et al.* (2007)]. Colloids are not occasional entities and in some sense most pharmaceutical products exist as at some point during their delivery to the body as colloidal systems.

Colloidal 'drugs' (nano-drugs; nano-medicines) and materials and those based on colloidal dispersions can be conveniently divided into *two* main groups, largely based on their make-up:

- Microscopic structures that are homogeneous/heterogeneous in nature:
  - Controlled release applications and bioimplants – crystalline aggregates (e.g. apatites), gel networks (microgels and nanogels) and encapsulation devices, enteric and other coatings, osteointegrants and scaffolds [Hsu *et al.* (1996)].
  - Microgel/microparticles/nanogels and solid mini-carriers for oral, buccal, nasal, gastroenteral delivery; they can be formed from polymeric surfactants or dendrimers to make 'effective pro-drugs', for use in localised targeting [Sarker (2006)].

- Sub-micron structures such as vesicular (liposomal) and micellar mini-carriers [Speiser (1998)]:
  - Liposomal preparations are used routinely for: enzymes, proteins, gene therapy (non-viral transfection), chemo-labile drugs. There is a range of liposome formats ranging from single bilayer (SUV) to multiple bilayers (MLV) [Speiser (1998); Hiemenz and Rajagopalan (1997)]. Liposomes routinely make use of phospholipids from natural sources such as soya. There are a number of drug-bearing liposomes such as AmBiosomes (containing amphotericin B), Daunoxomes (containing daunorubicin) and niosomes containing nonionic emulsifiers. Commercial versions of liposomes include Pulmozyme™ (encapsulated DNAase enzyme) for cystic fibrosis administered to the lungs, Taxosomes™ (paclitaxel, chemotherapeutic) and Doxil™ (doxorubicin, anti-neoplastic) which are chemotherapeutic liposomal preparations. Liposomal drug use dates back to 1959 and discovery by Bangham; however as a mainstream drug delivery form they were first proposed by Gregoriadis [Gregoriadis (1973)]. In recent years use of antioxidants, mixing of lipids [Nielsen and Gohla (1998)] and surface modification with PEG (poly(ethylene glycol)), antibodies and ligands has been considered [Valtcheva-Sarker *et al.* (2007)].
  - Microemulsions – using emulsifiers such as bile salts [Nielsen and Gohla (1998)] and often using a secondary emulsifier to create self-assembling, miniature, thermodynamically stable, clear emulsions ('nanoemulsions').
  - Micellar aggregates form spontaneously above a pre-defined concentration; they are used ubiquitously in pharmaceutical applications. Daktarin® a

micellar preparation, for example, contains the active micronazole (anti-fungal) housed in the core of the particle.

The products that make use of natural ingredients and exist in lyophilised form such as liposomes require additional use of preservatives, such as antioxidants, and cryoprotectants such as glycerol. These are incorporated as an essential feature of the colloidal drug systems that make use of emulsifiers and surfactants (lipids), and are critical excipients in terms of the purity, integrity and consistency of the raw materials. Essential quality control testing should always take account of variations in the purity and polydispersity of simple and polymeric surfactants, and source excipients from a reputable supplier. Nano-medicines, in addition to the two categories presented above, may also include antibodies (see Biopharmaceuticals, below), conjugates and clusters and chemically conjugated drugs which are represented in Figure 5.6.

Nano-medicines include a diffuse group of therapeutic agents [Schuster *et al.* (2006)] that are considered as the same group of therapeutics largely based around their size. They represent a group of medicines used for therapies as diverse as chemotherapy and cosmeceutical applications. However, they do, as mentioned earlier, represent a new growth area of pharmaceutical research patenting and use, often based on improved efficacy. It is estimated that is costs approximately one

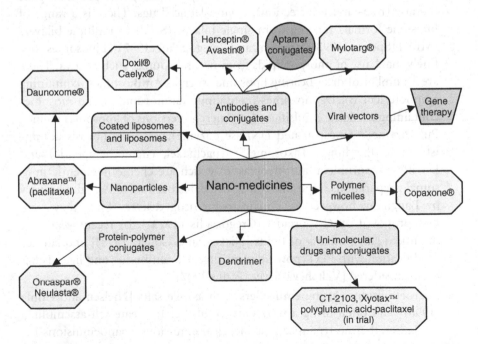

Figure 5.6 Representation of the various forms of nano-medicine. Adapted from [Duncan (2005); Torchilin (2001)]

billion Euro to develop each new chemical entity (new drug) and, as such, investment and laborious careful clinical testing mean the time involved in the physical realisation of a new product and a usable range of products can be quite large.

A brief illustration of the time-scale of nano-medicine productions is given below:

- 1997 – Rituxan (Rituximab) antibody; 2000 – Mylotarg antibody (conjugated to calicheamycin)

- 2001 – PEG-Intron (interferon)

- 2002 – Zevalin antibody (conjugated to Yttrium-90); 2003 – Bexxar (Tositumozab) antibody

- 2004 – Macugon (PEG-aptamer)

- 2004 – Avastin antibody.

Many of the newly devised nano-drugs are used widely in chemotherapy and although very significant steps have been made some failings are still apparent. This may be due to the over-zealous expectations of the pharmaceutical science communities that have been reaching out for new and improved means of specific cell targeting. Assessments of recent anticancer therapies are presented in Table 5.8.

There are *five* consistent themes of drug development associated with the production of new nano-medicinal entities and these are:

- The increasing use of 'smart' model materials and devices

- Developments pertaining to *ex vivo* diagnostics and portable applications, including therapies

- Aspects of therapy associated with imaging and the resolution of those images

- Therapeutics and focused delivery of drugs

Table 5.8 Details of many nano-medicines used for anticancer therapies since 1990

Vector	Device	Drawbacks
Natural	Protein carriers	Humanised antibody
	Antibodies	immunogenicity?
	Recombinant proteins	Release profile of
	Antibody-targeted	encapsulated drugs?
	liposomes*	Encapsulated
	Viral vectors	drugs – RES/MPS
		compatibility?

* – came from Ehrlich's 'magic bullet' concept in the early 1900s and supplemented by Bangham's 1960s concept of liposomes [Duncan (2005)].

- Themes associated with novel technology, in terms of production and translation from research to practice (scale-up) and the validation, regulation, safety and efficacy of new products and how this can be scrutinised effectively.

The improved efficacy over coarse suspensions and obvious interest in nano-drugs is associated with the nanoparticle size that facilitates easy passage [Fahmy *et al.* (2005); Sarker (2006); Valtcheva-Sarker *et al.* (2007)] into cell and sub-cellular compartments. Nano-drugs work well because on a crude level (temporarily disregarding active targeting mechanisms) they are of the same order of size as cell membranes (~8 nm) and much smaller than the target eukaryotic cell (~15 µm) or blood vessels and tissue, and are therefore afforded a degree of penetration.

These drugs also tend to work well because high loading of colloidal particles is possible (up to 70 per cent by volume) and the release profile can be strategically engineered by using antibodies and chemical ligands (see Table 5.9). Their value lies in the shielding and protection afforded to the drug itself and of the ability to access a wide range of tissues, for example to circumvent the blood-brain barrier (BBB) [Allen (2002); Torchilin (2001)]. In this manner nano-medicines are now used routinely in a number of cancer therapies.

Biopharmaceuticals

Drugs of this type represent a large category given the increased uses of biotechnologically produced hybridoma monoclonal antibody and recombinant technology, and the increased popularity of sourcing naturally occurring

Table 5.9 Chemotherapeutic nanoparticles and clinical trials

Ligand	System	Drug	Tested	Cellular Target
Folate	Liposomes	Doxorubicin	*In vivo*	Leukaemia cells
Albumin (Abraxane®)	Antibody–drug conjugate	Paclitaxel	*In vivo*	Breast cancer
Galactose	Poly(lactic acid), PLA	Retinoic acid	*In vitro*	Hepatocytes
Von Wildebrand factor (RexinG®)	Viral particles	Cyclin gene	*In vivo*	Pancreatic cancers
Integrin	Liposomes	Raf gene (cell signal)	*In vivo*	Melanoma cells
Fibrinogen	Albumin	Radio-isotopes	*In vivo*	Tumour vasculature
Aptamers	PLA	–	*In vitro*	Prostate epithelia

Aptamers are oligonucleotide fragments used in cell nuclear targeting [Fahmy *et al.* (2005); Sarker (2006)].

medicines from plant extracts [Bloomfield and Baird (1996); Walsh and Murphy (1999); Doblhoff-Dier and Bliem (1999); Johnson (2003)]. Stability and routine testing [FDA (1996); Walsh and Murphy (1999)] is an essential part of manufacture because the products tend to be inherently variable in potency and form. It has been estimated that by the end of the millennium the vast majority of drugs will be of the biopharmaceutical type, an estimate based on the recent changes in the number of diseases treated using biopharmaceutical products [Walsh and Murphy (1999), p1–34]. Table 5.10 gives an indication of the relative value of a number of specific and more generic medicines.

In addition to drug molecules the development of 'whole cells' to provide cancer vaccine immuno-therapy, such as Melacin® (in 2000), Canvaxin® for melanoma or Oncophage® for autologous tumour tissue [Schuster *et al.* (2006)] and now approved cytokine immunotherapy are some examples of the widespread growth of biotechnology in pharmacy.

In 1999 the global market for biopharma products was $7–8 billion and this was predicted to increase five-fold by 2005/6. Modern pharmaceutical sales are of the order of $200 billion with only about 1 per cent of pharmaceutical companies being biopharmaceutical producers *only*, as compared with the 10 000 pharmaceutical companies with more than 5000 products. The promise of 'biopharma' has yet to be seen in full. At the turn of the century and at the time of writing this book 50 biopharmaceuticals had gained regulatory approval. This can be considered to be small given the number of organisations working in the area but indicates the difficulty of approval and screening of new drugs. In North America and the European Union (EU) private companies represented

Table 5.10 Development trail of recent lead compound candidates and promising biopharmaceutical products

Active pharmaceutical ingredient	For use in	Approval period/time	Value (x $million)
Digitalis	Heart conditions	1900	?
Quinine	Malaria		
Pecacuanha	Dysentery		
Mercury (non-biopharmaceutical)	Syphilis		
Insulin	Diabetes mellitus	1982	1000
Human growth hormone	Growth deficiency	1985	?
Interferons, α, β, γ	Cancer, anti-viral, multiple sclerosis	1986–1990	1000
Erythropoeitin	Anaemia	1989	2000
Interleukin-2 (Proleukin-cytokine)	Cancer	1992	50
PEG-Aptamers	Gene therapy	2004*	?

* – not fully approved for routine use.

almost 85 per cent of global organisations, with the biggest share coming from the EU (46 per cent) [Wiley-VCH and FDA. (2006)]. Of the 4500 or so global companies 82 per cent were private companies and this indicates the importance of specialised or 'niche' manufacturers in this area. Many are technology transfer bodies from academic institutions working on recombinant DNA technologies and hybridoma antibody proteins with more than 350 NCEs currently under clinical trials making use of gene therapy and anti-sense technology.

The major quality considerations for biopharmaceutical and biotechnology products used as, and in, therapeutics are contamination, purity and chemical consistency. One of the major concerns for manufacturers of such products is downstream processing and clean-up [Graumann and Premstaller (2006)] of the product. Products such as recombinant therapeutic proteins that might include hormones, therapeutic toxins and interferons are particularly hard to screen because of their microbial origins. The fermentation process conditions and mode of manufacture have a bearing on process yields. The use of complexing agents and chromatographic resins for purification purposes and the extent of purification can add to the cost and thereby impair the viability of the drug. Proteins are often precipitated using chemotropic agents, then using refolding buffer to reconfigure the protein. Obviously, the protein (5 per cent purity is needed as a starting point for biopharma products) must be obtained in a configuration that will work as a drug. Problems of protein-based biopharmaceuticals can include proteolysis and loss of form (or even toxicity) and since downstream processing usually represents more than half of the cost of a biopharmaceutical this part of the process must be considered and validated at great depth. Routine manufacture makes use of on-line monitoring processes that might include UV, pH, and conductivity sensing using process analytical technologies (PAT) as recommended by the FDA. The demand of process flexibility sought by the developer, yield and sheer quantity of drug required can complicate matters.

Key quality indices for the products of microbial and cell-line biotechnology normally include counts and if required the absence of the following:

- DNA (as an impurity)
- Pyrogens
- Viral particles
- Microbes and cell fragments (and lysis products)
- Proteins (as an impurity).

Since very large proportions of all biopharmaceuticals are dosed and delivered in solution form as either parenteral, topical, or pulmonary medicines, the removal to the extent that defines product ideality of these components can pose a real problem to the manufacturer. A list of current biotechnology products is provided in Table 5.11.

Table 5.11 Non-exhaustive list of biopharmaceuticals currently in use

Products	Sub-class	Example	Relevant details and concerns
Blood derivatives	Blood clotting factors I–V; VII–XIII	VIII/IX–haemophilia	Infectious agents and pathogenicity e.g. CJD prions, hepatitis A–C particles, parasites?
	Anticoagulants	Heparin/Hirudin (from leeches)	
	Thrombolytics	Streptokinase Streptococcus spp.	
Proteins	Serum proteins	Human serum albumin Bovine serum albumin	Infectious agents (as above)
Therapeutic enzymes	Metalloproteins	Zn/Cu superoxide dismutases and Zn-proteinases – chemotherapy and anti-inflammatory use	
	Asparaginase	Leukaemia	
	Digestive enzymes	Pepsin, lactase	
	Debriding enzymes	Collagenase, trypsin	
	Recombinant enzymes	DNAase (Pulmozyme™) – cystic fibrosis[a]	[a]Genetech, 1993
		Glucocerebrosidase	Gaucher's disease
Recombinant hormone therapies	Insulin	Humulin[b] Humalog – quick acting analog	[b]Eli Lilly, 1982
	Human growth hormone	Protropin, 1985	
	Gonadotrophin	Gonal F (fertility treatment)[c]	[c]1996

(continued overleaf)

Table 5.11   (*continued*)

Products	Sub-class	Example	Relevant details and concerns
Haemopoietic growth factors	Red blood cell stimulating factor (replacement for usual rate10^6 cells/second)	AIDS, cancer	
Interferon[d] and interleukins	Cytokines – induce resistance	[d]Roferon-A, 1986	[d]Roche for leukaemia and hepatatis
	Growth factor for T lymphocytes	Proleukin, 1992	
Vaccines	Hepatitis surface antigen BCG antigens	Infanrix, Recombivax	Viroid inactivation extent?
Monoclonal antibodies	B-lymphocyte- immortal cell hybridoma	Mylotarg[e]	[e]Celltech, 2000
Gene therapy agents	Antisense technology interference with mRNA	Haemophilia, cancer, AIDS, hypercholes- terolaemia	Specificity, toxicity

Shown are some of the diverse uses and developmental augmentation in biotechnological pharmaceuticals from the 1990s [Walsh and Murphy (1999); Wiley-VCH and FDA (2006); Schuster *et al.* (2006)]. Superscripts a–e simply highlight connections within the data presented in the table, usually pointing to a time and organisational involvement.

Biological products and products derived from blood carry additional risks over and above those of the form of the active drug. With plasma derivatives there is the additional risk of inclusion of HIV or hepatitis viruses and also the risk of CJD prions. Some of the main producers of biopharmaceuticals for the UK market include GSK, Wyeth (Ireland), Aventis, Chiron Vaccines, Eli-Lilly (Liverpool) and Avecia (oligonucleotides), although the global list is considerably greater (see Table 5.11). One biopharmaceutical, paclitaxel (Taxol™), used in anti-cancer therapy and extracted from yew tree, that is administered as sterile USP injection accounts for 22 per cent of cancer therapeutics in the UK.

To control biotechnology product quality the manufacturer has to consider the extent of incorporation of toxic substances (pyrogens) and the form of the molecule. In the UK (equivalent bodies exist elsewhere) this is in part achieved with the assistance and guidelines provided by the MHRA's Committee for Proprietary Medicinal Products (CPMP). This body provides the manufacturer

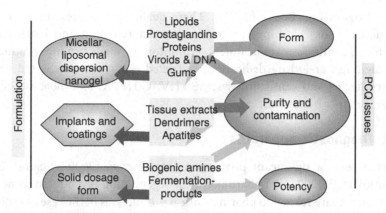

Figure 5.7 Biologically derived medicines and formulation, purity, consistency and intrinsic quality concerns

with basic instruction on the provision and maintenance of quality systems to ensure the safest product (Figure 5.7). The basic outline of requirements is:

- Suitable staff training that might include additional 'biologics'-specific training and both record-keeping and precautionary measures for the immunological status of staff.

- In-process controls are soundly in place that permit the full evaluation of starting material suitability, with the aid of preparation protocols and a highly regimented and extensive system of checks to confirm finished goods purity, consistency and quality (PCQ) [Walsh and Murphy (1999); Doblhoff Dier and Bliem (1999); Sarker (2004)].

The problems of producing high PCQ-grade biomaterials, biopharmaceuticals and biotechnology products include:

- The poor consistency of biological material

- The difficulties in obtaining (ultimate goal) a high degree of QA

- Product uniformity from batch-to-batch

- Lyophilisation and preservation of the chemical moiety and loss of functionality or potency.

All of these must be validated to give the manufacturer an indication of the risk involved in producing the medicine in any significant and useful quantity. In terms of the manufacturing environment a biopharmaceutical facility should have sterilisation equipment, containment cleaning (CIP/WIP and SIP), water-for-injection (WFI), automation, lab support and purpose-built quality systems over and above routine validation in an ideal manufacturing facility [Slater

(1999); Vogleer and Boekx (2003)]. These quality procedures should include clean-down (room cleaning) protocols, product retention and batch control procedures, continuous environmental monitoring, single product productions (campaigning), area/drug dedication, restricted movement, dedicated heating and air filtering and ventilation systems (HVAC/HEPA) and bioactive effluent segregation or containment.

5.2.2 Diagnostic apparatus

This represents a significant proportion [Skoog *et al.* (2000); Eggins (2003); Howbrook *et al.* (2003)] of pharmaceutical organisation revenue. In some cases the diagnostics are related to pharmaceutical products, as in the case of radio and magnetic resonance pharmaceuticals and biopharmaceuticals. Typical diagnostic products that might soon find their way into community pharmacies rather than the clinical environment include sensors for blood urea, oxygen, carbon dioxide, ions (such as calcium, sodium and potassium), glucose, cholesterol, and specific enzyme tests. The link to clinical assessment makes control over the products and maintenance of quality a significant issue.

5.3 Quality assurance aspects of medical gases, devices and miscellaneous product manufacture

Medical gases are medicines as they feature in the British (BP) and European Pharmacopoeias (EP) and the British National Formulary (BNF); but more than that they are used in hospitals and pre-packed products (e.g. aspirators) for administration of medicinal compounds (see also anaesthetics, Section 5.3.3). They are used in a clinical or surgical environment by piping through mixer manifolds or use of compressed gas cylinders as the gas source, and used in environments as diverse as operating theatres, intensive care units, imaging departments such as radiology and on the hospital ward. All cylinders must be clearly identifiable as to their content since errors could prove to be fatal. Each cylinder is given a colour code and the cap is also colour coded depending on the content, for example:

- The gas cylinder itself – nitrous oxide is blue, carbon dioxide is mid-green, helium is brown, helium-oxygen mixture is black and helium-oxygen-nitrogen mixture is light green.

- The cap or end – nitrous oxide is blue, carbon dioxide is green, helium is brown, helium-oxygen mixture is brown-white and helium-oxygen-nitrogen mixture is pink.

Medical gases include oxygen, nitrogen, carbon dioxide, nitrous oxide (laughing gas), helium, nitric oxide and a variety of binary and three-component gas mixtures. Any gas cylinder or source must also carry an expiry date, description of content, a filling date and an identification or batch number.

Other risks involve contamination of the gas source and leakage containment, particularly relevant to noxious gases such as carbon dioxide, and faulty pressure leading to dosing errors – particularly relevant to anaesthesia. A well-planned layout is essential as the pressurised gas can represent a real danger to staff safety. Any alterations or modifications need a permit to work and appropriate authorisation. Staff training in use, bulk ordering, stock control, cylinder storage, auditing and incident reporting and actioning remain the responsibility of the qualified/responsible person.

Sterile intravenous (*iv*) medicine units are used in the hospital but controlled by the hospital pharmacy QA department. They are prepared *in situ*, usually by nurses and clinicians, having been purchased from an external commercial supplier. This does not negate the need for regular routine QC practices pertaining to the raw materials. In fact there is a risk of contamination by foreign matter, alien pharmaceuticals and microbes (moulds, Pseudomonads *Staphylococci, Streptococci*, Coliforms), and a risk of faulty preparation in terms of dosage, dilution, precipitation of the active and administration to the patient. The units are purchased for use in chemotherapy, antibiotic administration, cardio-pulmonary therapy and in some cases for parenteral nutrition and use of vitaminised products, but have the advantage of being largely pre-prepared in a licensed production unit, which is subject to scrutiny by official regulators. Products of this variety are manufactured in an environment of isolation, laminar flow cabinets, aseptic filling and specialised containment suites (see Section 5.2.1). The principal risks of most products are those already mentioned but additionally ones of cytotoxicity to formulator and recipient and a notion of whether the product is sterile or effectively sterile (aseptic). The products, if bearing licensed medicines, are subject to the Medicines Act 1968 in the UK. If unlicensed then batch size is limited and the shelf life of the product is also substantially limited by a product exemption clause in the same Act; this local regulation is likely to apply universally, not just in the UK.

Most *iv* products are normally presented in five formats, 'polyfusor', bottle, bag, vial or ampoule and these are made of a range of materials ranging from plastic copolymers, glass and PVC to polyethylene so that QC tests would follow the product integrity and breaches. Consequently, routine quality control would also examine proximate analysis, microbial loads, photo- and chemo-stability (e.g. retinoids), pH (buffer presence), chemical interactions resulting from mixing, and the product temperature profile.

A number of other products such as extruded solid dispersions of drug, solid powder injection (such as those prepared by Powderject UK), surgical and internal implants and novel medicaments such as smart capsules are all subject to appropriate rigorous QC tests which define their suitability for intended use.

Multi-component models require uniformity of content tests. The regulation of such products depends on their status as aids or as medicines and the appropriate pharmaceutical license. This group can also include imaging solutions and suspensions.

5.3.1 Multi-component packaged medicinal products, including inhalers

Nebulisers and medical aerosols are subject to routine ISO physical and mechanical tests relevant to their use. In the case of pre-prepared injections (as used for allergen shock and diabetic insulin shots), these would require active content tests and those pertinent to product purity. The practical working of the equipment requires key outlined physical and mechanical tests for unit integrity and conformance to specifications.

5.3.2 Radiopharmaceuticals and imaging agents

There are two aspects to nuclear clinical medicine and radiopharmaceutical use: therapy and diagnostics. Top industrial manufacturers and representatives in the medical diagnostics market include Amersham Health, Schering and Bristol-Myers Squibb; Tyco Healthcare and Medical Imaging between them having the vast majority of the market share. Radiopharmaceutical application provides information about degenerative disease such as stroke, dementia, coronary artery disease and cancer but also goes hand-in-hand with anatomical imaging such as positron emission tomography (PET), magnetic resonance imaging (MRI), ultrasound (US), computed tomography (CT), optical imaging and planar radiography. CT is one of the most widespread imaging techniques for medical diagnostics, but generally uses X-ray not gamma-ray energy photons. Radiopharmaceutical imaging always requires the use of a radio-labelled substrate or medical diagnostic product, which binds to structures within and accumulates in the anatomical features being studied. In the case of gamma-ray isotopes the image is obtained with the aid of a gamma camera and often using positron-emitting radiopharmaceuticals (PERs) and in this case the product group includes metabolites such as ^{18}F-fluorodeoxyglucose (^{18}F-FDG), a potential metabolite used readily by tumour cells.

A radiopharmaceutical is 'any relevant medicinal product, which when ready for use contains one or more radio-nuclides included for a medicinal purpose.' These radiopharmaceuticals are prepared from a precursor, as in the case of technetium (99metastable, ^{99m}Tc), which is prepared from Molybdenum (^{99}Mo) in a solvent extraction generator. Calibrations of radiochemical and radiopharmaceutical purity and potency in the US are based on measurements at the National Institute of Science and Technology (NIST) or equivalent designated institutions in the country concerned. Activity measurements for the gamma-ray-emitting radio-nuclides are made using a specialised liquid scintillation spectrometry and ionisation chamber. The calibration process also includes identification of radio-nuclidic impurities by germanium spectrometry. Radio-nuclides are usually produced in a nuclear reactor, generator device, or most commonly in a cyclotron.

There are a number of important considerations when using radio-nuclides (isotopes); these include the nature of the particulate emitted radiation, its energy (20 keV to a maximum 600 keV) and thus penetration power. Tissue penetration and the half-life ($t_{1/2}$) of the isotope determine medicinal product shelf life and handling danger [Burns (1978)]. Ideally $t_{1/2}$ should be of the order of one hour to one year for routine radiopharmaceutical uses, with most products having a value in the day, week or month range. Radio-nuclide emissions are of three principal types:

- α-particle (helium atom): penetration and subsequent cell damage is very localised

- β-particle (electron, β^- or positron, β^+): tissue damage is localised

- γ-particle (gamma ray): considerable tissue and organ penetrating power.

The energy of isotope emissions diminishes as a general rule in the order $\gamma > \beta > \alpha$ radiation, with the lattermost being the 'weakest'. There are many radio-nuclides (Table 5.12) but some do not possess ideal properties for use in medical diagnostics and these include hydrogen, sulphur and phosphorus isotopes.

Diagnostic medicinal applications, such as for PET applications allow a dose that is usually up to a maximum of 400 MBq/day (0.01 Ci) for a person. Other considerations should of course be the physical properties of the isotope. However, therapeutic medicinal applications frequently require dilution of pre-formulated medicines and mean they are often administered by intravenous (*iv*) format via a sterile pre-fabricated pack or a syringe. In this case this administration means the radio-nuclide is not covered by the Medicines Act 1968 [UKRG (2006); Amersham plc (2006)] but suitable QA activities should be in place to ensure patient safety. Guidance in the UK comes from the Department of Health via the Administration of Radioactive Substances Advisory Committee (ARSAC).

The most commonly used isotope is technetium (^{99m}Tc) and is followed by pertechnate (^{99m}TcO$_4$). This isotope is favoured because it has known impurity levels (when compared to formal standards e.g. FDA, USP). It also has mono-energetic emission, a short half-life, a high photon yield and can be generated in the clinic via generator technology using solvent extraction from ^{99}Mo, which is its most common impurity of about 0.1 per cent. Prostate therapy uses radioactive iodine (I^{125}) or palladium (Pd^{103}) seeds that are implanted to irradiate the tumour.

Table 5.12 Non-exhaustive list of commonly used radio-nuclides

Isotope(s)	Half-life ($t_{1/2}$)	Principal photonic energy gamma emission	Electron (β^-) emission energy (MeV)	Positron (β^+) emission energy (MeV)
Chromium (Cr) 51	Month	M	M	–
Cobalt (Co) 57/60	1/2 year/ 6 years	M	M	–
Fluorine (F) 18	Few hours	H	L	H
Gallium (Ga) 67/68	Few days/day	M (many)	–	VH
Indium (In) 111/113m	Few days/few hours	M/H	M	–
Iodine (I) 123/125/ 131	1/2 day/ 1/2 day/ week	M/L	L	–
Thallium (Tl) 201	Few days	M	–	–
Technetium (Tc) 99m	Few hours	M	–	–
Oxygen (O) 15	Few minutes	–	–	VH
Nitrogen (N) 13	Few minutes	–	–	VH
Carbon (C) 11	Few minutes	–	–	VH
Energy emission key: low (L) <100 keV; moderate (M) 100–300 keV; high (H) 300–600 keV; very high (VH) 1–2 MeV				

Other products include strontium[89], which is used to target bone cancer, and iodine[131] for thyroid tumour treatment [Sarker (2006); Amersham plc (2006)].

Current state-of-the-art radiopharmaceuticals and radio-labelling method diagnostics are often based on 'natural' molecules, such as antisense oligonucleotide fragments (aptamers), small mono- or oligosaccharides, proteins or hormones, antibodies and biogenic peptides [Owunwanne et al. (1995), 1–99].

Magneto-pharmaceutical colloids using surfactant and polymer-stabilised systems and ferromagnetic solutions or suspensions form the bases of many preparations used for imaging purposes. Imaging agents include Lipidol® (Laboratoire Guerbet), which is a radiological contrast agent formulated as an O/W emulsion using iodinated poppy seed oil for use in external beam radiotherapy.

Table 5.13 Some radiopharmaceutical products and *iv* administration

Radiopharmaceutical example	Diluent(s) used	Comments
^{99m}Tc kits	1 % w/v sodium chloride	Dextran, citrate, polymer pro-drug conjugates
^{111}In Octreoscan®	1 % w/v sodium chloride	Instructions provided
^{123}I human serum albumin, (Isopharma)®	1 % w/v sodium chloride; 1 % benzyl alcohol or albumin 15 mg/ml	Details in product guide
^{59}Fe Ferric chloride	0.6 % w/v sodium chloride; 1 % sodium citrate dihydrate	Precipitation possible
^{51}Cr Edetate (Cr EDTA)	1 % w/v sodium chloride; 1 % benzyl alcohol	
^{90}Y Yttrium silicate	Not recommended for dilution	Colloid must be maintained at high pH

The meaning of all chemical symbols has been mentioned in an earlier table.

Table 5.14 Solution, *iv* and colloidal radiopharmaceuticals (see also *Colloidal mixtures and biomedical nanotechnology*, Section 5.2.1)

A selection of various complex radiopharmaceuticals	
Colloid-suspension format	Solution format
^{99m}Tc-serum albumin and latex microspheres	^{99m}TcO$_4$ (aluminium ion, usual impurity <0.1 %)
^{99m}Tc-sulphur or tin colloid	^{75}Se-selenomethionine
^{99m}Tc-erythrocytes	^{125}I-iodocholesterol
^{125}I-fibrinogen	^{125}In-quinolinepolyphenol and other heterocyclic compounds
^{198}Au gold colloid	
Aerosols and macro-aggregates of proteins	

The oil is retained more specifically by cancerous liver cells and can also be conjugated to specific chemotherapeutics for treatment such as using doxorubicin, epirubicin and 5-fluorouracil. Barium sulphate dispersions which form biomedical imaging agents (barium meal, barium enema formats but also perfluorooctyl-bromide (Imagent) and specific contrast imaging agents (see emulsions and colloids)) also find a key medical role.

Routine QA and QC activities [Oropesa *et al.* (2002); Nakao *et al.* (2006)] and clinical uses of radio-isotopes come under the following regulations and guidelines over and above various relevant specific regulation in the UK, US and Euro-zone countries. Also pertinent in this case are The Medicines Act, 1968 and The Misuse of Drugs Act, 1971, specific to the UK. The essential skeleton guidelines for 'radiopharma' are provided by:

- US Food and Drug Administration, 21 CFR Parts 315 and 601 (RIN 0910-AB52) – Regulations for *In Vivo* Radiopharmaceuticals Used for Diagnosis and Monitoring

- UK The Medicines for Human Use (Marketing, Authorisations, etc.) Regulations 1994, Schedule 1, Regulation 3(2) – covering general indications

- EU Article 16 of Council Directive 75/319/EEC – covering general indications

- UK Medicines (Administration of Radioactive Substances) Regulations 1978 (Regulation 2) – clinical practice.

Given the risk associated with product and concerns among routine practising clinicians, certain product details are required for official purposes in any host institution; these include [European Parliament (2006)]:

- The source of product

- The person to who supply was made

- The quantity of supply

- The batch number

- Details of any suspected adverse reaction to the product.

Good quality clinical testing and routine QC should mean both suppliers and hospitals are required to accurately calculate doses (chemical potency), preparation and preparation in-house for use. Additional biological tests should include specificity (perhaps done as part of validation) and biologic activity of the radiopharmaceuticals administered, associated with efficacy, that is an appreciation of the labelling of drugs with isotope or chemical label efficiency. Standard QC tests would include particle-size measurements for colloidal preparations, limit tests and half-life calculations by way of estimating the shelf life and its compliance to the United States Pharmacopeia (USP) or BP or EP. Other chemical tests would include the product safety (sterility, apyrogenicity, non-antigenicity, 'non-toxicity'), purity and overall quality. The department that administrates the radiopharmacy practice would need to show its appropriateness in terms of centralised organisation, its daily planned operations, handling, operator shielding, isolation/containment, personal exposure monitoring (lithium chloride indicator test), safe waste disposal and quality control procedures in place [Hesselwood (1990); UK HSE (1985); ICRP (1977)].

5.3.3 Anaesthetics

Anaesthetics are usually applied to the skin, stomach, rectum, intravenously, intra-muscularly, into the spinal cord or to the lungs. They act by a variety of mechanisms; general anaesthetics often act on ion-channels at a molecular level. They range from complex heterocyclic compounds to monatomic gases. Anaesthetic chemicals and classical pharmaceuticals fall into three distinct classes depending on administration; these are those responsible for general anaesthesia, systemic (injection) anaesthesia and local anaesthesia [Whalen *et al.* (2005); Conzen (2005); Vickers *et al.* (1991); Parker (2006); Hopley and van Schalkwyk (2006)].

General anaesthetics include volatile and non-volatile varieties; they instigate a reversible loss in consciousness, which can result in light or deep anaesthesia (where muscle control is lost). Volatile anaesthetics include the non-flammable fluorinated varieties desflurane and sevoflurane (most used), the halogenated varieties halothane (fluothane) and isoflurane, chlorinated varieties such as trichloroethylene and chloroform [Whalen *et al.* (2005)] and other varieties such as nitrous oxide, carbon dioxide (see medical gases), ether and xenon. These volatile anaesthetics are usually mixed with gases such as nitrogen, oxygen, air, and argon (see Section 5.3) for the purposes of correct dose administration and altering the physico-chemical properties of the gas. Where these are liquid they must first be vaporised, most having boiling points less than $60°C$, before application. Volatile anaesthetics are combined with gases stored under immense pressure, $50-140 \, kg/cm^2$ (~ 130 bar) and as such the cylinder, valve, flow meter and vaporiser must work effectively to give the correct dose. Non-volatile general anaesthetics include two short-acting barbiturates, thiopentone and methohexitone, and the often-used sedative etomidate. These are usually administered in a 1 per cent solution and to a dose of approximately 100 mg.

Systemic anaesthetics are administered by injection or *iv*-line and include opiates, codeine phosphate, pethidine, paracetamol and non-steroidal anti-inflammatory drugs (NSAIDs) such as ibuprofen and indomethacin. The most common intravenous anaesthetics act as central nervous system depressants and include propofol and the barbiturate thiopentone, etomidate and ketamine hydrochloride (HCl). Premedication uses sedative to ease the transition and these use tranquillisers metochlopramide (Maxolon) and perphenazine (Fentazin), and the neuroleptic sedatives droperidol (Droleptan) and the benzodiazepines, diazepam (Valium), nitrazepam (Mogadon) and midazolam. Many of the systemic drugs come under the Medicines Act 1968 and the Misuse of Drugs Act 1971 (Schedule 2) in the UK and are therefore controlled substances. The third type of anaesthetic is the local anaesthetic applied by injection or topically;

these include two basic sub-types: basic esters – tetracaine, benzocaine, procaine HCl, cocaine, and basic amides – lidocaine and ropivacaine. They are frequently administered as hydrochloride salts to augment their solubility, and work by reversible inhibition of nerve signalling.

Areas of particular concern in terms of QA and QC include correct dosing for special preparations in the dispensary relevant to hospital pharmacy. Additional dangers of using compressed gas, the metal corrosion of 'moist' halogenated volatile anaesthetics e.g. halothane, and the toxicity of substances to the habitual workers mean an environment which needs regular checks. Some of the substances are known to show renal, hepatic and alveolar irritation and toxicity and so should be used with appropriate ventilation.

6

Process control via numerical means

Decision making in following the findings of a quality control laboratory or as part of following the routine production of materials is difficult at the best of times. The job of the qualified person is difficult enough but can be assisted by using statistical assessment by way of provision of a rationale to help the decision-making process. The use of statistical process control (SPC) provides a vehicle for monitoring the 'quality' level of a process by application of statistical methods in all stages of production. In most cases it is based on a normal frequency distribution (NFD) and problems of bias and kurtosis that relate to the normal distribution [Johnson (2003); Snee (1990); Snee (1986); Buncher and Tsay (1994)] illustrated in Figure 6.1.

The 6-sigma format represents a model for the best of current practices in routine manufacture because six standard deviations (sigma, s, σ) should really capture the entire process because random fluctuations in a process might account for as much as 1.5 sigma. Most manufacturing considers that the sample 'outliers' (outlying values) should be captured within three standard deviations of the sample theoretical average (μ). The 6-sigma method came to the forefront of business practice from the engineering industries and can be used to define process capability. In a standard everyday process most of the samples (~ 70 per cent) would be caught within plus or minus one standard deviation of the process average.

This can be observed in the format shown when process bias produces kurtosis (skewed distribution) or when sample number (n) is scrutinised as part of total lot size (N); and the ratio used for any study is always crucial. The standard deviation (σ; s – gives an indication of intrinsic variability or process precision) and the average (μ, x_{bar}) give an indication of relative systematic variation and

Quality Systems and Controls for Pharmaceuticals   D K Sarker
© 2008 John Wiley & Sons, Ltd

The Normal Frequency Distribution
– Gaussian symmetry

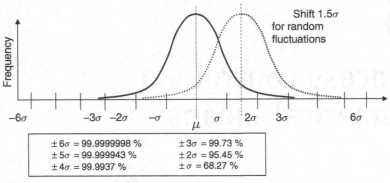

$\pm 6\sigma$ = 99.9999998 %	$\pm 3\sigma$ = 99.73 %
$\pm 5\sigma$ = 99.999943 %	$\pm 2\sigma$ = 95.45 %
$\pm 4\sigma$ = 99.9937 %	$\pm \sigma$ = 68.27 %

- 6-sigma $\cong 2\cdot10^{-7}$ % defects
- 3-sigma $\cong 2.7\cdot10^{-1}$ % defects

Figure 6.1   6-sigma (6σ) processes and their relevant modelling in the normal (Gaussian) distribution of data

- **When is a NFD not appropriate?**
  - – Bias introduced?

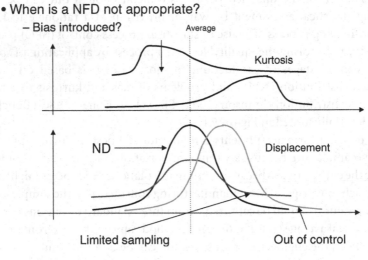

Figure 6.2   Observed modification, misrepresentation, bias and kurtosis seen with poorly sampled data during validation and statistical process control (SPC)

process trueness, respectively, from the theoretical true value (represented by a Greek symbol). The true value is referred to in estimations of the accuracy [Harris (1999); Buncher and Tsay (1994)] of any processing procedure (Figure 6.2).

The primary aim of SPC for pharmaceutical manufacture is to define the inter-relationship of factors within a process, to give an appreciation of the probability of an event and thus provide a means for improved evaluation and to indicate the significance of an event, and thus permit tighter control over the loss of quality (see Chapter 3). SPC also provides a yardstick by which to assist essential decision-making, evaluate chance variation, designate relevant 'assignable cause' to an event, decide on the validity of data outliers and follow data trends. Finally, use of statistics is valuable in terms of process validation because it allows the manufacturer to decide on the permanence or scope for improvement in the patterns established within a process.

6.1 Charting and quality inspection

Many processes are evaluated by reference to so-called quality control (QC) charts. These charts form one of the fundamental cornerstones of routine inspection on any ongoing process and may relate to clinical evaluation, routine production and many other applications. The standard format of a quality control chart for the relative humidity of the processing suite is shown in Figure 6.3. The variable parameter evaluated can change, however it does usually include a feature of the process that is critical to the production of the best quality product. This parameter is usually monitored as a function of the stage in a process, for example the start middle and end of the run or various batches spaced evenly throughout the production run or campaign. The aim is to keep the product (as measured) between the lines of the upper and lower acceptable limits, usually represented by a spread of six standard deviation values (a normal distribution of data).

Figure 6.4 shows how the QC chart fits in with a notion of limit specification and the 3 or 6-sigma process. A normal distribution of data (NFD) is superimposed on the standard plot that shows the usual cut off point at plus or minus three sigma (99.73 per cent) values but as part of a 6-sigma process may be extended yet further to include 99.9999998 per cent of all samples. The Figure illustrates how most of the measured sample variation should lie within the boundaries of a NFD. An occasional outlier may represent either a point of warning meriting corrective action (likely) or a random fluctuation (relatively

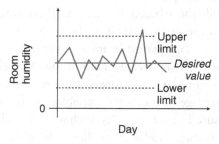

Figure 6.3 A generalised Shewhart-style quality control chart for a 6-sigma process showing the environmental humidity conditions and limits

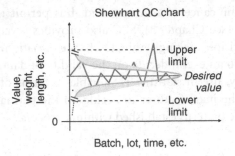

Figure 6.4 A Simplified representation of a quality control (QC) chart that is found in routine pharmaceutical manufacture and in a laboratory or clinical context [Loftus and Nash (1984), 1–70; Carleton and Agalloco (1999), 1–45; Cundell (2004)]

unlikely). The *centre line* is the average or nominal value (x_{bar}; x_{db}) of the product that may be taken from product specifications (desired value). The construction of quality control charts is based on three additional specified limits based on multiples of the standard deviation for the measured product; these are:

- Upper action level (UAL) or lower action level (LAL), that is the centre line ± 3 standard deviations

- Upper warning level (UWL) or lower warning level (LWL), that is the centre line ± 2 standard deviations

- Upper control level (UCL) or lower control level (LCL) that is the centre line ± 1 standard deviation.

QC charts can be based on *attributes*, for example that might cover whether a product possesses the appropriate quality (e.g. works or does not work), conformance, etc. These are usually related to routine inspections and make use of P/F charts (pass or fail charts are often based on complex mathematical functions). As a rule these are used less routinely than charts based on variables, which will form the major part of the subject matter on charting. The second variety of charts include those based on *variables*; these describe the character of a product or index of the process, such as the weight, or hardness of a tablet. Charts based around variable quantities are used ubiquitously as part of assuring and controlling quality and to provide an indication of developing or established trends. There are four fundamental forms of these charts. The first variety is based on either an average of measured variables (x-type chart) or an average of the range in measured variables (R-type chart). The advantage in the latter is that both average (mean, modal or median, *Mo* or *Me* values, respectively) and relative variability are expressed within the chart. X-type charts make use of a quality which represents the average of averages (double bar mean or x_{db});

this provides the chart with currency which takes into account occasional drift that might be a consequence of many factors. The other three forms of charts are represented by the Shewhart type, moving average type and cumulative sum (CuSum) types. Averages are always based on the number of samples (n) in the original sample; the upper (UCL) and lower control limits (LCL) are set according to three average standard deviation (s) variations of the averaged sample data. These values are 'tweaked' for the purpose by application of a range of constants that can be obtained from tables and whose value depends on value of the *sample* size. Likewise with R-type charts again the upper and lower control limit values are set to three sigma and corrected for sample size with various constants, yet again obtained from official inspection tables.

Shewhart charts are subtle in their difference from standard 'x' and 'R' charts. They are based on having a value measured against time and are characterised by having a target value (based on established validation) and experimental standard deviation which sets the control level. Moving average charts are Shewhart-type charts but the average and thus the target are based on the four most recent measurements of the sample. Most QC charts fall in line with the central limit theorem (CLT), which in synopsis says that as the sample number approaches very high or infinite values, corresponding to the lot size, then the experimental average (x_{bar}) and standard deviation (s) approach the theoretical values (μ and σ, respectively).

CuSum charts can be constructed using two methods: by experiment or from mathematical treatment of validation data. The charts use a basic graph and a transparent movable v-shaped mask (v-mask) that can be moved from sampling point to sampling point. The charts plot the incremental change between the current and previous sample (Figure 6.5a). They in very clear form (Figure 6.5b) show the trend in the data. If the process data follow a projected angle (θ) exactly then this represents a target value for a process to be in control. A data set that falls outside of the V-mask trend-line falls out of specification and indicates a point of attention and critical corrective action. This type of chart was popular in the past because of the ease of use and lack of necessity for mathematical skills to establish process failures.

Most routine control variable charts are based on a normal distribution (NFD) of data just as might be expected of process-measured quantities – tablet length and tablet hardness, etc. The Gaussian (bell-shaped symmetry) form of the plot is characterised by the positioning and magnitude of sample value (x), theoretical mean (μ) and standard deviation (σ). The NFD forms the basis of x-charts but not the basis for R-charts. Attribute-based QC charts are more frequently based on the concept of failure (non-compliance, non-conformance) and defective units and as such are based on either the binomial distribution (probability or P-charts) or the Poisson distribution, which characterises the ratio of defective units and forms the basis of C and U charts. P-type charts show the proportion

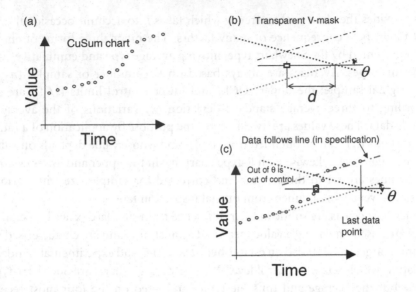

Figure 6.5 The V-mask format of a cumulative sum (CuSum) chart

of unacceptable units plotted against batch, set against a centreline representing average defective ratios and upper and lower limits (roughly three times the average number of defectives multiplied by a constant factor) and are usually used for random inspections with a fixed number of samples. The Poisson-related charts (C and U type) are characterised for large sample numbers where the number of defectives is constant.

Quality level and a notion of inherent variability are represented in control charts and this is of use in determining the suitability of a given process of manufacture. Four general scenarios exist and are represented phenomenologically in Figure 6.6. The centreline represents the values that should be seen most often and the shaded area represents the margins and scope for variation (standard deviation) within the arrowed vertical limit lines. The first example (a) represents the ideal case and of course is seldom seen in practice. Versions (b) and (c) are the next best examples with case (b) being more favoured because of constant quality and type (d) represents the worst of all possible cases in which variability and quality (trueness to the desired value or centre-line) are never achieved.

These descriptions of quality pertain to a normal frequency (NFD) and increasing use is being made of even stricter process control where a 6-sigma format is applied [Kolarik (1995); Pyzdek (2001); Deshpande (1998); ITSM (2005); Six Sigma Tutorial (2005)]. The so-called 3-sigma control levels are now an international standard for capturing quality. According to a coarse but useful description, a 3-sigma process should encompass 99.73 per cent of samples and permit approximately 2.7×10^{-1} per cent defects, whereas a 6-sigma process should encompass 99.9999998 per cent of samples and permit approximately

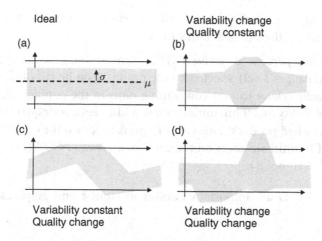

Figure 6.6 Common defects in representation of data on a QC chart. The inherent variability and data quality may change with time for a number of reasons

2×10^{-7} per cent defects; rather notably the risk of injurious product, 'escaping' detection is reduced in a 6-sigma product.

The 6-sigma process referred to earlier (Figure 6.4) is based on a lack of tolerance for product failures. The philosophy uses five drivers to ensure compliance and these are *definition* of process or product goals, *measurement* based on using a key comparative standard, *analysis* of inter-relationships and causality, *improvements* and optimisation of the results, and finally *control* of the process through suitable measuring and appropriate corrective actions [Pyzdek (2001); Deshpande (1998); ITSM (2005); Six Sigma Tutorial (2005); Johnson (2003)]. These are often referred to as DMAIC indices and are useful in ensuring a process is customarily 'highly' compliant. The approach also uses the notion of *process capability* and this is a numerical index where the difference between upper and lower limits (UAL – LAL) is divided by six sigma for the process.

Process quality may also be defined in reference to the ratio of defective to non-defective produced materials. This grouping forms the basis of a number of important conceptual indices linked to notions of quality at various stages of a process that are of significance in statistical sampling programmes and approaches; these are:

- Acceptable quality level (AQL), an arbitrary value selected by the producer that usually relates to non-conforming units in the population, associated with a probability of rejecting perfectly good materials that is called the producers' error (this probability is given the symbol α).

- The inverse of low-level failures is higher level, and gives rise to the notion of defining a process in terms of unacceptable quality level (UQL). This is generally only used where product failure is likely and thus numbers are high.

- Average outgoing quality level (AOQL), which pertains to the quality of material leaving the site of manufacture.

- Lot tolerance percent defective (LTPD), an arbitrary 'worst quality' level (perceived danger level) selected by the producer on behalf of the customer, which usually relates to non-conforming units in the population, associated with a probability that the manufacturer would accept a poor quality/dangerous or otherwise bad product. This type of mistake is called the consumers' error, in light of its ultimate effect on the consumer [Snee (1990)] (this probability is given the symbol β).

AQL and LTPD are important notions in tabular and graphical sampling plans.

6.2 Sampling plans

The basis of most common sampling plans that represent a guide to acceptance or rejection of a batch of produced medicines is the establishment of critical specified values. This is achieved with reference to establishment of appropriate hypotheses. In general, the AQL would be accepted if the product was less than or equal to a pre-specified value and likewise the UQL would be rejected if a value determined was greater than or equal to a pre-specified value. The probability and patient-related consequences for making an erroneous judgement coupled to the financial penury for the institution are critical considerations. Probabilities are encapsulated in notions of the producers' (α, false alarm raised) and consumers' (β, failure to detect flawed product) errors. A given process is considered to run in a stable form when the proportion of defectives is as stated (less than or equal to the AQL). The specified AQL must be set at a level which is reasonably obtained otherwise there is a risk of an unacceptable and unsustainable number of product failures.

There are four main types of statistical sampling plan as opposed to physical sampling that involve the mechanical action of extracting a portion. These are based on selection of a pre-set number of samples (n) and thereby give a degree of certainty and assurance in the results obtained that they are in line with the size of the sampling. As a general 'rule-of-thumb' greater sample sizes come with a greater assurance and a more effective representation of variation within the sample population. The sampling regimes are:

- Single ($n = 10$), with the designated number of samples taken once.

- Double – in the first instance $n = 6$; on failure (inferred uncertainty), an additional 6 must be taken.

- Multiple, where initially 4 samples are taken, then a further 4 and a final 2; here supplementary testing occurs after initial failure.

- An alternative method, given by sequential sampling – here notions of a handicap value and penalty values are used to give a score when factored into the number of defects found. Below a predetermined value or score the batch is rejected, but it allows for sampling to be undertaken in stages and variable quantities of sample to be taken at each sampling point, which directly influences the magnitude of the penalty and handicap.

Various formats of sampling plan exist ranging from quality statement or code of practice that offers some idea of sampling strategy, to graphical representations such as the widely used operating characteristic curve (OCC) and Dodge and Römig Model (pass, re-test, fail zones within a plot), through to tabular sampling plans. These provide an effective check on quality by permitting a batch of materials to be assessed and subjected to a consistent evaluation. There is little scope other than in the code of practice for differences in interpretation that provide an essential tool for the qualified person. The standard representation of an OCC [Kolarik (1995); MHRA (2002)] is illustrated in Figure 6.7. Notably quality is defined by the number of defective materials present, AQL and LTPD values are represented, and the probability of acceptance of a batch can be made based on the position of the quality on the sigmoid-form curve.

The ideal representation of an OCC given a theoretically perfect number of samplings ($n = 100$ per cent) is given in Figure 6.8. In reality no such form ever exists other than when an infinite and impractical number of samples (total lot size, N) are measured. In this case the chances of making an erroneous judgement are obviously zero.

As the number of samples interrogated is increased the OCC assumes a narrowing and consequent tightening of the degree of confidence in measured samples and their extension to the size of the lot. This occurs primarily because a better representation of the sample population is established and a trend in the behaviour is easier to establish (Figure 6.9). Where the sample size is ten the equivalent probability is 0.93, however when the sample number is raised to 50

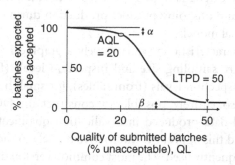

Figure 6.7 Simplified representation of an operating characteristic curve (OCC), which is a graphical sampling plan

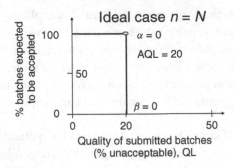

Figure 6.8 Simplified representation of a theoretically perfect sample number ($n = N$ or 100 per cent testing) for an operating characteristic curve

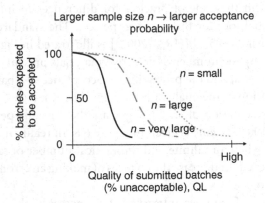

Figure 6.9 Simplified representation of an operating characteristic curve showing how the form of the OCC changes with a low or high number of individual samples (n) taken

the probability is raised to 0.98 (98 per cent). OCCs are usually constructed so that 60 per cent or so of the samples will be expected to be accepted at 2AQL. There are two types of OCCs, type-A that relate to small lots, and type-B that relate to large lots and on-going product production during sampling and that often follow binomial models.

An operating characteristic curve is merely a graphical representation of a sampling plan where sampling size and inspection levels (from ISO or other tables) or special inspection levels (from tables), for small batches for example, are given. It is also entirely possible to construct an OCC from 'scratch' using experimental data produced in a validation qualification, and previous production has used this approach. Alternative models to the graphical form of an OCC are also frequently used. The most common version used is one shown in generalised form in Figure 6.10. This tabular form shows sample size, acceptance and rejection criteria and the corresponding AQL value, and the appropriate

Single sampling plans - for normal inspections

Sample size code	Sample size (n)	AQL					
		0.01					1000
		Ac	Re	...	...	...	
A	2						
...	...						
...	...		*x* *y*				
...	...						
...	...						
R	2000						

Use first sampling plan below arrow

Values increase left to right and top to bottom

Use first sampling plan above arrow

If sample size (n) is lot size (N) do 100% inspection

Figure 6.10   Simplified representation of a tabular sampling plan for attribute testing

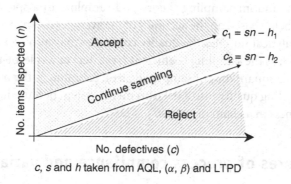

$$c_1 = sn - h_1$$
$$c_2 = sn - h_2$$

Accept

Continue sampling

Reject

No. items inspected (n)

No. defectives (c)

c, s and h taken from AQL, (α, β) and LTPD

Figure 6.11   Sampling plan alternative to a conventional OCC, proposed by Dodge and Römig

guidance instructions. Many operators find the tabular form of the OCC easier to use than the graphical representation.

In one rarely-used graphical example according to Dodge and Römig (Figure 6.11), the representation demonstrates to the user the values that are relevant to sample rejection, continued sampling (and possible re-testing) and acceptance. This is based on a positioning of the product quality in terms of the number of non-conforming units against the number of objects that should be inspected. The plot represents an alternative but has not been widely used because of the familiarity and widespread usage of the OCC format.

When considering physical sampling the most common concern is that the portion represents an appropriate description of the lot. In a similar manner with statistical sampling the sample numbers examined must represent an effective description (of the status, compliance, etc.) of the large numbers that are produced routinely during a production campaign in an industrial setting. In terms of a NFD this can be defined in terms of the confidence levels for a particular sample size. These are a function of the average (μ), standard deviation (s), t-test constant and sample number (n), according to the formula:

$$\mu = (ts)/\sqrt{n}. \qquad (6.1)$$

Here, t values are taken from t-test tables and are normally given as 1.96 for a 95 per cent confidence limit. This sampling indicator is an important aid as it can be used to give an estimate of acceptable sampling number as a rule of thumb to move away from 100 per cent testing. It can be re-configured to give the sample number required (n) with respect to experimental uncertainty or error (which is the average value less bias; $\mu - \mu_1$).

$$n = [(ts)/(\text{error})]^2. \qquad (6.2)$$

Other indications of suitable sample size can be generated from sampling size guides such as random sampling theory and sampling to a specified accuracy, with the latter being based on the accuracy of the average (mean, median or modal value). For analytical samples the quality controller may also consider sampling parameters such as the sampling coefficient (see later) or a rule-of-thumb sample number (e.g. the square root of sample lot size). Sampling is the most important aspect of instituting quality into a measurements system because it represents the point of weakness in a chain of events.

6.3 Measures of process compliance and variation

The accuracy and variability of data are measures of location and data dispersion and give an overall impression of process compliance. Typical indices include the average (mean), or modal average or median values (less frequently used), standard deviation and variance and derived quantities based on these such as relative standard deviation, coefficient of variation or standard error. Use of the mean and standard deviation can give an indication of the error of a measurement (see Figure 6.1). These simple quantities can also be used along with sample numbers to model differences within a process via statistical functions such as F-tests. An additional valuable notion is one of process error, where the overall error of a measurement is a composite of bias introduced into a measurement by (a) faulty equipment or (b) systematic faulty methodology and the additional feature of the statistical (random) error. A number of factors can be involved in the error relating to a measurement but more often than not poor GLP or GMP practice has a significant role to play. Errors and measurement bias may also be subject to propagation with misdiagnoses (relating to GCP) or mis-recording (relating to GLP) of the examined individual components that constitute the final product or process. Errors of this type can have a devastating impact on decisions that are made because they can be additive (less effect) or more unfavourably multiplicative (significant impact) with respect to one another.

Statistical analysis and simple mean and variation recording have a number of key roles within the evaluation of the suitability of a process. SPC provides a means of:

- Probabilistic modelling to aid in current or future validation based on the ability to undertake limited sampling

- Assessing whether a process is out of specification (set limits) and therefore worth continuing with

- Assessing whether a process is changing with time, with respect to its output and by extension is thus influenced by variations in the process conditions in use at the time of manufacture

- Provision of a simple guide to the status (pass or fail) of the product batch

- Giving a criteria and weighting to the decision to pass or fail the batch

- Evaluating the adequacy of the sampling in terms of having taken enough samples (n) to ensure acceptable confidence in the results [Loftus and Nash (1984); Carleton and Agalloco (1999)]

- Judging the relative risk of passing the product and the consequences of making an erroneous decision.

There is not a single routine manufacturing procedure that takes place during mass manufacture that does not take account of the decisions and actions listed above. As such the central guarding role of statistical analysis as a guide to assist the qualified person (QP) or laboratory manager make decisions is assured.

7

Product verification and the role of qualified personnel

Effective QA has a number of fundamental requirements that are usually orchestrated by a quality circle that often utilises the skills of the QP or other responsible person in the organisation. The minimum requirements in this case are those of a process having a clear audit trail. ALL alterations to standard procedure must be reported and explained with any anomalous behaviour being recorded (as this will help in further qualification). All procedures should use unambiguous step methodology where 'the working' SOP should be verified by the QP and all instrumentation or processes requiring calibration should have appropriate regular certification. An appropriate paper trail for control or record keeping should demonstrate a linearity of the process; this avoids backtracking and role confusion leading to possibilities for adulteration or cross-contamination of the product. Documentation should be based on systems which permit a hierarchical form of limited access to prevent security breaches in process control software, batch records and SOPs.

An essential part of document and record generation is safe archiving, and just as retention samples are kept for chemical confirmation then QC documentation must also be secured and stored safely so that the records may be audited at a later point. The constituents of documentation used for quality assurance purposes are based on three sub-types of records or guides and these are documents based on policy or practice, reference documents, and second generation derived documents, such as results and conclusion or summary reports. The various classes of documentation and their component elements are represented in Figure 7.1.

The sophistication of controlling documentation is implicit in the effective working of a QA department and the TQMS activities of any pharmaceutical

Quality Systems and Controls for Pharmaceuticals   D K Sarker
© 2008 John Wiley & Sons, Ltd

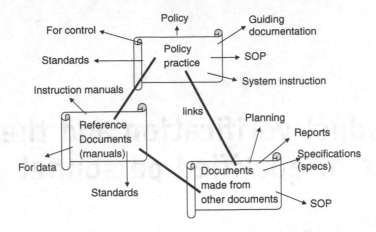

Figure 7.1   Model representation of quality assured documentation

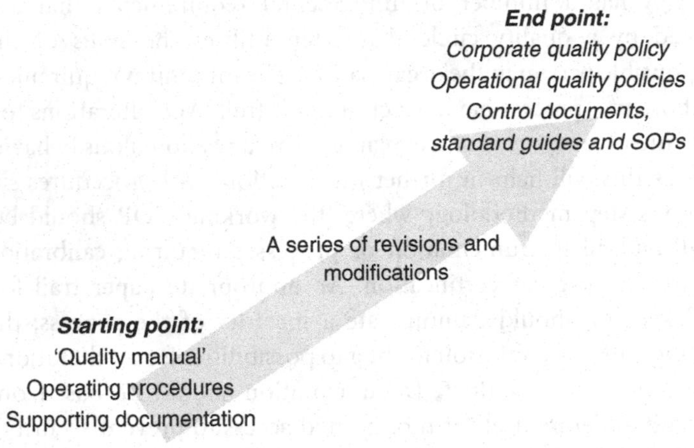

End point:
Corporate quality policy
Operational quality policies
Control documents,
standard guides and SOPs

A series of revisions and
modifications

Starting point:
'Quality manual'
Operating procedures
Supporting documentation

Figure 7.2   Documentation levels of specificity and evolution to purpose with time

manufacture because documentation usually provides the key instructions for efficient departmental composition and functions. Documentation should evolve (Figure 7.2) to fit the environment required and the pre-requisite needs of the department concerned. The starting point for a new process may be loosely built around a quality manual and basic SOPs but should be enhanced to provide a vision of corporate quality policy (standards maintenance) control, operating policies and standard guides in a better QA model. Documentation should be integrated with good overall practice (GXP) and is crucial in terms of current GCP, GMP and GLP. Robust and in-depth record keeping and reporting is pivotal to clear passage through regulatory inspection and auditing.

7.1 Batch documentation

Batch records and documents are used to describe the changes to the product as it passes from one unit operation to another. Documents of this type are produced to a template and completed by the personnel in charge of production, with a final sign-off by both the production manger and the QP. In a standard form used for example in tablet manufacture they may include reconciliation weights between starting materials and the product at a particular stage of completion. Typically this may include *in* (for received goods) and *out* weights, tests on the sample (usually 10 dosage forms) and a series of routine and standard in-process tests that are matched against the required properties of the drug system. For tablets these might include the unit hardness, unit weight, unit dimensions, unit appearance and colour coding, presence of the tablet break-bar, organisational stamp or drug identifier. Similar product indices are applied to topical and parenteral dosage forms and help to ensure that the product is consistent.

Suitable batch documentation should detail the product form and dosage strength along with a document number, version, and the number of pages in totality, issue number, details of relevant SOPs at the point of use and identification of author, and authoriser. There are four basic types of batch documents and these relate to incoming materials, production (including coating and packaging), finished product, and quality control. For quality control documents good QA practice would routinely expect details of batch number (BN), date of manufacture, details of SOP (mechanical tests, bioavailability tests, assay and impurity tests), key analytical tests and numbers of samples indicated, with appropriate descriptions of the sample average and variability and an indication of the pass or fail status of the product indicated unambiguously. Each page must contain document details, references and details of guidance and regulation essential [Hoyle (2006); MHRA (2002)] to the evaluation. Any anomalies or 'routine' deviations to the standard process of manufacture (product license or stipulations of manufacture) or QC testing must be evaluated on a change control form (Table 7.1). The document itself should be referenced to the parent document and must undergo a series of peer reviews culminating in the authorisation of the most responsible person, normally the QP or QP/managing director.

The change control document must have a consensus across the team and of the most responsible person. In some case revisions to the information represented in the document may be sought before approval.

Table 7.1 A specimen of an idealised change control document, required for minor and major revisions to a validated process

PROCESS CHANGE CONTROL

Document no. ___XXX_____

Revision no.   __XXX_____

Required for   ___product XXX_ dosage strength X_

Requests from   ___person(s) concerned about changes_____

Reason for change to
Process   ___ elaborate_____

Requestor   ___ Mr J Bloggs, role_____

Signatories terminating with QP (circle appropriate action):

	Accept		
1_____	yes	change	no
2_____	yes	change	no
3_____	yes	change	no
4_____	yes	change	no

Comments

e.g. process details, faults and problem, remedial action, regulatory implications, impact on the product

7.2 Standard operating procedures

Standard operating procedures are pharmaceutical or practice-based methods (protocols) often referred to as SOPs. Normally they assume two positions based on their evolution for a new process or improvement to an existing process. At first a 'working copy' (valid protocol) of the SOP is established; this is easily identified as it carries details of the application, group under which the SOP applies, carries a valid-from and valid-to date mark and bears the signature of two or more responsible persons, one of which is normally a qualified person. Secondly, an SOP can be in circulation, as part of continual improvement or validation, which may be modified subject to verification for general use by QA. This improved version is not for normal use but may bear the latest modifications and will subsume the position of a valid protocol after the relevant QA approval, usually via change control documentation. The SOP details all knowledge that is crucial to undertaking a unit operation from start to finish and it may also occasionally include extra information of value in fault-finding or for troubleshooting.

7.3 Guides, overviews and validation plans

Successful and unfaltering regular routine manufacture arises from the creation of a validation plan (VP) or validation master plan (VMP) and associated report (VMR) that represents a basis for everyone in the team knowing what, where, why and how to undertake key procedural steps, and a team effort to produce a medicinal product of the best quality. In principal, a VP is nothing more than a collection of guidelines and apportioned procedural steps united with a basic explanation of the key steps and their significance. The aim of such documentation is to avoid confusion and the wastage associated with re-working materials or the scrapping of batches based on clumsy preparation and the lack of dutiful rigour in QC testing.

Amalgamation of the most significant aspects of production such as GMP, validation, authority regulation and internal inspection, and software or laboratory support within a purpose-built manufacturing environment, when given appropriate qualification and validation, stops process uncertainties influencing the facility (Figure 7.3). In this manner having such systems in-place prevents the failure of the site or production area to produce high quality materials and therefore constitutes an essential part of the total control of quality (TQC) concept. In addition to TQC the process conformance of a product or manufacturer is controlled externally via a number of regulatory bodies. The activities of the regulators are broken down into those of legality covering human medicines and other activities that a manufacturer might be involved in such as manufacture of veterinary medicines. The legal aspects of control relate to drug product efficacy, safety, training and education and internal management of resources and records within the regulatory organisation. This is shown in Table 7.2 for the European Medicines Evaluation Authority (EMEA) that controls medicine and medical device output within the Euro-zone area.

The major considerations of on-site practice that fall under the remit of the regulatory body include the site VMP, individual product VMP, conformance to the organisation's product or site license and conformance to the overriding law (in the UK this would be the Medicines Act 1968). However, other regulatory bodies such as the FDA or MHRA, and even local guidance, have to be considered in the formulation of a 'generic' code of practice. This code of practice is the template by which all processes of drug manufacture and working standard operating procedures (wSOPs) are put in place. Consequently it becomes clear that having the appropriate sectional considerations (Table 7.2) leads to tighter control and better organisational efficiency.

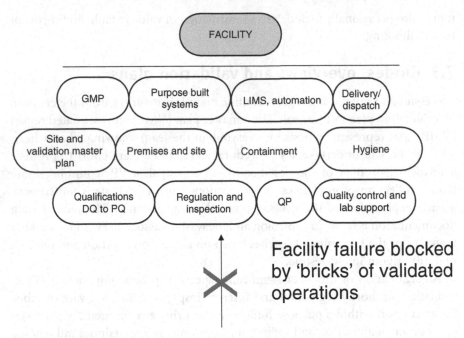

Figure 7.3   Planned facility and key site components within the framework of regulated manufacture

Table 7.2   The regulators of medicines and developed pharmaceutical products in the UK [Case (2006); Tambuyzer (2002); Schacter (2006), 114–270; Rang (2006), 255–298; MHRA and Roche (2006); EMEA (2006)]

EMEA directorate and executive control areas				
Legal, management and audit teams				Animal medicines
Administration	Communication	Human medicines: pre-authority	Human medicines: post-authority	
		Considerations		
Organisation + finances	Informatics + project + conference + document management	Miscellaneous drugs + safety/efficacy + quality	Regulation + safety/efficacy + medical information processing	Inspection + safety + marketing

7.4 The duties of the qualified person

The qualified person (QP) can in principle represent any person in an organisation with appropriate skills and knowledge. The key word in the last sentence being 'appropriate', since the knowledge base to make safe and judicious assessments best representing the organisation carries immense responsibility and an element of risk [MHRA (2002)]. A fuller description of these requirements for UK and European QPs can be found in EU directive 2001/20/EC article 13 (Section 3.3.2). Equivalent professional standards are applied certainly across Europe and the developed world and even into the more tightly regulated countries within the developing world. The role of a QP has undergone a transformation and this is largely as a result of test-case lawsuits in recent decades and a greater degree of moral and organisational responsibility. Up until 1977 a company could use staff with institution-deemed satisfactory experience, membership of an affiliated professional organisation (and a reliance on professional integrity) and know-how. However, this led to inconsistencies in the depth of the knowledge base and a lack of 'regulation of the regulators'. Since 1985 and in Europe, validation in terms of formal professional accreditation has been needed and this is based on conformance to EU directives EC 75/319 and EC 81/85, aligned with pharmaceutically relevant professional membership (e.g. Institute of Biology, Royal Pharmaceutical Society of GB or Royal Society of Chemistry, in the UK) and a post-graduate qualification.

The role of a QP is a complex one (shown in part in Figure 3.7) involving liaison between the various elements of a TQMS. At some level the QP is generally involved in most of the key decision-making with regard to product suitability. However, their primary objective is to check that the manufacturer's batch (produced product) exactly matches the manufacturer's or product licence (legally defined permitted product) within very tight limits of variability. This is to ensure that the product meets with the appropriate sample under which clinical testing was performed and for which there is substantial proof of safety. The QP ensures compliance with certification of the satisfactory nature of the product and a risk evaluation based on a comprehensive understanding of the knock-on effect of product inconsistency and non-conformance.

The responsibilities of the QP are threefold: to the customer (most important), to the profession represented, and finally to the organisation. This responsibility and accountability to oneself, professional and peer groups and most importantly statutes and the law are enough to prevent clashes between ethics and financial pressures such as the need for the highest productivity. To work most effectively a QP must be fully conversant with aspects of:

- The law, QA administration and management policies
- Process and chemo-metric statistics and QC practice
- Fundamental concepts associated with biology, toxicology, microbiology and hygiene
- Fundamental concepts associated with analytical sciences, medicinal chemistry and pharmaceutics (product formulation sciences)

- Fundamental concepts associated with pharmacology and drug action
- Engineering considerations associated with laboratory and factory organisation and production processes, including packaging and the properties of materials.

These elements relate to the production of routinely safe medicines and medicinal products. On a routine basis the most essential duties of a QP include conformance to the various product licences, that GMP is followed through appropriate validation of manufacturing and routine process test points (such as a check on production conditions). Duties may also include confirmation that there is FDA (or other relevant body) approval of process deviations with additional sampling deemed necessary if the process or product does not assure consistency or quality but also that quality documents are endorsed appropriately and that internal (self-inspection) and external audits are coordinated and are legally compliant.

8
In-process and on-process QC testing and control

A number of key activities are used routinely in quality control and these include physical sampling and recovery (see lot sampling in manufacture, Section 4.2.1), with confidence and a reproducible sample recovery as the primary goal, and the use of validated methods for QC activities. All laboratory tests should be performed using valid analytical methods (VAMs) and supporting chemometric evaluation. Newer processes also make use of on-line (ancillary to the process) or in-line (*in situ*) testing that forms an essential part of the equipment. Many successful in-process devices include process analytical technologies (PATs), where the analysis is taken out of the QC lab (off-line testing) and placed within the production environment.

In-process evaluation is now favoured for the QC of a number of specialised products [Sharp (2000); Doblhoff-Dier and Bliem (1999) Muller *et al.* (1996); Bloomfield and Baird (1996)] such as parenteral medicines. These products are customarily variable and carry with them a risk of cross-contamination or microbial degradation and because they exist in colloidal or solution form they are subject to higher rates of chemical degradation. Examples of products assessed in-process using PATs include:

- Pulmozyme™, enzyme-carrying liposomes in sterile liquid

- Taxosomes™, paclitaxel liposomes in sterile liquid

- Doxil™, doxorubicin liposomes in sterile liquid.

The sterility needed for these therapeutic products requires *in situ* testing and application of the highest standards of freedom from pathogens. In this case such

Quality Systems and Controls for Pharmaceuticals D K Sarker
© 2008 John Wiley & Sons, Ltd

quality tools as 6-sigma standard production are still relevant to safe medicine production but with 2×10^{-7} per cent non-conformance this still poses an unacceptable risk in mass manufacture. Since 6-sigma encapsulates 99.9999998 per cent of the product this can be combined within a hurdle approach of heat (one in a million to billion risk) and non-heat sterilisation methods to reduce the risk [Pyzdek (2001); Cundell (2004); Johnson (2003)]. Even batches of produced medicines can be evaluated at the point of manufacture by using non-continuous methods but based on robotised and automated PAT rapid-screening systems.

8.1 Process analytical technologies

Process analytical technologies are those, usually of an analytical or pharmaceutics nature, which are used to describe an aspect or product or part-product of a process. They are subject to an appropriate SOP. Examples include weight, hardness, exhaust vapour analysis (GC-MS), drug and moisture content. The latter two may be measured by Fourier transform near-infrared (FT-NIR, IR) methods and the former-most by the resonant frequency of a granulate using acoustic sensing [Ansel *et al.* (1999); Norris and Baker (2003)]. In addition to on- and in-line sensing, PAT methods may also evaluate QC samples in an off-line format following a large production run if taking and measuring samples might interfere with the production run. Other favoured methods that form the basis of PAT methods include attenuated total internal reflectance (ATR) FT-NIR for liquid samples, ultra-violet, fluorescence or visible spectro-photometric opacity measurements, and electrochemical sensors for microbiological testing (impedance or conductivity methods). These are often linked to an appropriate statistical processing package to give an estimate of the validity of the result. As part of using such a system considerable validation of the test method would be required. In one such system a multiple PAT approach was used for assessment of end-of-processing for tablet granulate formed in a fluidised-bed drier, using a microwave method to assess water content, and dynamic light scattering to evaluate particle size; this was combined with multivariate statistical evaluation of the data to indicate the end of the processing regime [Reich (2005); Holm-Nielsen *et al.* (2006); Halstensen *et al.* (2006); Hausman *et al.* (2005)].

8.1.1 On-line sensors

Routine in- and on-process tests can be incorporated into a standard piece of equipment that runs in continuous or semi-continuous (semi-batch) mode, to discriminate suitable and unsuitable production materials. The on-line system can also be used with a batch production of medicines. Common methodologies include the use of a 'magic eye' that might use metal detection, colour changes and light obscuration to monitor changes in samples. This methodology is used in many instances where ampoule-based products are produced. Supplementary methods may make use of suitable 'smart' materials that respond to temperature, pH or other changes and thereby indicate that suitable processing has taken place. Colour-based heat-sensitive tapes are now used routinely for thermally treated products. An increasing use of microchip smart technology is to record the position of a tamperproof-labelled (usually contained in the packaging) product during processing and distribution. Finally, microbiological safety can be assessed by using a number of tests such as impedance testing and the direct epi-fluorescence technique (DEFT) method that reflect cell numbers and viable cells, respectively. Such on-line testing reduces the need for 100 per cent testing, samples to be manually taken to the QC laboratory, which may not be on the site or the same part of the site, and in the most practical sense, laborious manual sorting.

8.1.2 Hyphenation in manufacture and process diagnostics

Small-scale manufacturing may use chromatographic analysis (HPLC, GC), including headspace analysis, but may also include relatively 'rarer' techniques for routine use such as mass-spectrometry and NMR. For biopharmaceutical products analysis may involve HPLC linked to MS; this provides the advantage of simultaneous quantification and identification on the same parent sample. Larger-scale manufacturing of pharmacist-only medicines (POM) and over the counter (OTC), in addition to special products, are more likely to use routine measurements such as hardness testing, total organic carbon (TOC), infrared (IR) and spectrometry or refractive index measurements. In recent years an increase in hyphenated methodologies has seen IR linked to differential scanning calorimetry (DSC) and acoustic technologies. Hyphenated analytical (and PAT) technologies of this type are used ever more because of the advantages of consistent sampling.

8.2 Analytical validation and clinical test validation (CTV)

The start-point for an analytical procedure that might be used as part of a clinical evaluation (e.g. metabolite or toxin assay) or for following a production process (e.g. in-process samples) is to decide on the aims and uses of the method. Validation of a test method is based in the initial research stage of forming a plan based on current practice and the objectives of the test, which is then used to make a suitable method (Figure 8.1a). Transfer of the method from an R&D environment to the working environment of a QC lab requires adaptation of the technique and conditions. Any methodology can be used in one of five characterisation respects with regard to a sample or a process. Analytical techniques provide information on identity, content and therefore purity, the

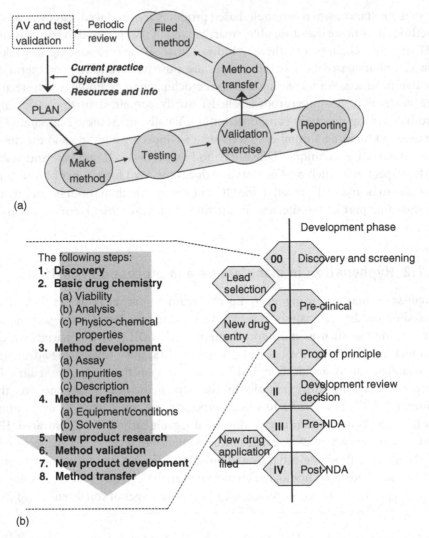

(a)

(b)

Figure 8.1 The analytical validation lifecycle made simple. The starting point for any activities is a strategic plan. Part (a) involves the analytical validation lifecycle and part (b) involves methodological evolution with respect to the drug development process

ease and extent of drug dissolution, the mechanical and physical properties and characteristics and the extent of degradation and the number of impurities present. The basis for assessment following the development of a new chemical entity that crosses from a lead (investigational) drug to that of one in commercial use is presented in Figure 8.1b. Most of the analytical validation effort is focussed between phase 00 (discovery) and phase I (pre-clinical testing) but should be in a suitably robust form to permit periodic review as part of further modifications in the filed method congruous with the product commercialisation scale-up (phase IV) process.

The validation itself should consider a comprehensive regime of tests based on the URS covering the DQ, IQ, OQ and PQ [Slater (1999)] as with manufacturing validation (see Chapter 4). Here, the performance qualification should also include basic proficiency testing that makes use of systematic parallel testing, internal and external auditing and the appropriate challenge testing. For an automated HPLC system, for example, this may involve the number and frequency of samples run that stretches the equipment to the extremes of customary use. As with other forms of validation, use of a continual-improvement cycle (quality control spiral) is integral to obtaining the best analytical method.

Validation of an analytical method should always consider the compliance of the method with established methods (by way of a safety check), the historical background (Figure 8.1a), method definitions, assay validation, and method development and alteration to make the method fit-for-purpose in the case of a NDA [Munden et al. (2002); Cledera-Castro et al. (2006)]. Even after the point of establishing a method that works well this does not mark the end of a process but merely a point of ongoing validation that might reflect ways of working such as inter- and intra-lab method transfers that are common with globally manufactured products. Aspects associated with networking of laboratory computers, lab information management systems (LIMS) and interpretive technologies [Neville (2005)] also need validation and are tied to the analytical method.

The 'fitness-for-purpose' of methodology and its significance lies at the heart of method selection and should form the foundation of the URS. The concept considers the use of a specific analytical procedure for the exact measurement of a *specifically* targeted analyte in a fully purposeful and methodical manner (Figure 8.2) that covers the sample selection, procedural steps, quality concerns

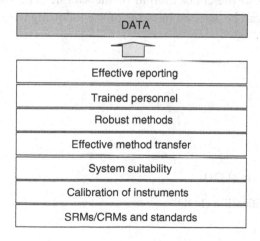

Figure 8.2 The standard components of analytical validation

and suitability indices. Appropriate methods should encompass reliability, the links between product suitability, with processing via testing (and customer assurances), and the final check of product safety. Validation of these methods may be undertaken piecemeal in a bottom-up approach or in terms of the validity of the end result as a top-down approach. All analytical validation efforts should make use of reliable system suitability tests that may have originated at the method transfer stage and have ties with the initial drug research and development [Selkirk (1998); Norris and Baker (2003)]. The highest quality data that is produced comes from an amalgam of correct reporting, correct methods, the suitability of a test system for the measurement, the use of appropriate reference standards (CRM and SRM), checked equipment and highly trained technical staff.

In order to maximise the best working and form of analysis, the general considerations of the technique should also include standardisation of the usage conditions, sampling regimes (heterogeneity in the sample) and its practicality; sampling aids, such as the sampling constant (weight needing to be taken) and the choice of sample using statistics (e.g. the Horowitz Trumpet) that defines sample size suitability [Harris (1999); Pritchard (1995)] are often helpful. The Horowitz trumpet approach is based on the relationship between the sample size (or mass tested) and the inherent variability in the results as determined by the standard deviation or coefficient of variation (COV), with smaller samples showing more variation. Careful selection and optimisation of the sampling conditions (e.g. making sure the sample number is correct, see Section 6.2) are important because they relate to information quality and the extent of quality assurances and decisions made on the product.

Additional considerations should include instrument calibration, control experiments and blanks and the appropriate recovery of ALL the analyte from the sample matrix that might be achieved best through chemical derivatisation, extraction and pre-concentration of the sample. The appropriateness of equipment used can be assessed in terms of the following ten system suitability indices:

- Accuracy

- Precision

- Linearity

- Range

- Limit of detection (LOD)

- Limit of quantification (LOQ)

- Sensitivity

- Specificity

- Selectivity

- Method robustness.

The most important parameters for most forms of analysis and for the analyst in terms of analytical instrument qualification are method *accuracy*, *precision* and *specificity*. Accuracy is often referred to as 'trueness' and uses an SRM with comparison by two methods. Accuracy is then determined using a chemically 'spiked' blank sample. Precision refers to the extent of agreement within a data set for not less than 10 repeats. Elements of precision include the method ruggedness, which accounts for intra-lab variation under subtle changes to experiment conditions, and also include the method reproducibility, which is the inter-lab precision often captured in terms of the coefficient of variation (COV), for which various formulae have been derived. We can further consider the precision of a method in terms of data and three levels of complexity: repeatability (consistency), intermediate precision (variations within the laboratory) and finally reproducibility (variations between different laboratories) when performed on the same sample.

Estimates of the relationship, for example in terms of linearity (linear regression, $y = mx + 0$) as determined on a conventional calibration plot (calibration curve), should show a correlation coefficient (R^2) of at least 0.99 over not less than 50 per cent of the data range. The plot should cover at least five data points, each of which should be based on at least several ($n \geq 3$) replicate measurements. Range is a parameter that refers to the extent of acceptable concentration with regards to linearity, accuracy and precision. In many applications the lower limit of detection and confidence in the measured value are important, such as for degradation products. These aspects are encompassed in notions of the limit of detection (LOD) that is represented by lowest quantity of analyte that is significantly different from blank (defined as the measured value for the blank plus three standard deviations), and the limit of quantification (LOQ) defined as the reporting limit (measured value for the blank plus ten standard deviations) where no analyte is detected. Sensitivity within a technique can be seen as the degree of sample dilution that still yields a reliable and sensible result.

Another practically useful index, in addition to accuracy and precision, is the method specificity, which is sometimes also referred to as resolution. It represents the ability to distinguish analytes from the matrix components, for example that might be represented by the baseline resolution in chromatograms. The value is confirmed by a test approach that challenges the equipment with respect to matrix composition, temperature and other key influential factors. The selectivity of a method is a useful indicator of the 'appropriateness' of a method because it shows the degree of freedom (as the selectivity coefficient, k) of the technique from interference. This selectivity can be relatively narrow as in the case of ion selective electrodes (membrane permeability) or broad, as in the case of UV spectrometry. The selectivity of a technique can also be defined in terms of the equipment output, for example with high performance liquid chromatography (HPLC) the selectivity is captured in terms of the peak retention time, number of theoretical plates and peak tailing. Conversely, with spectroscopy there may be poor selectivity as represented by peak width, peak overlap and signal overlap [Skoog *et al.* (1998); Buncher and Tsay (1994); Ahuja and Scypinski (2001)].

When a QC manager is adopting a new method transferred from R&D, an essential consideration for further adaptation of the method is the method robustness. Here the manufacturer's equipment tests are combined in a standard-ised but new test environment and the method is subjected to methodological variation such as alterations to the experimental pH, ionic strength, injection volume, flow rate, temperature and test duration. The analyst roles and the impact these might have on, for example a 'standard' packed HPLC column [Dejaegher and Heyden (2006)], or an alternative such as a monolithic column, can become significant to the experimental outcome, such as detection of an analyte. This might then be expected to have a bearing on the test output and its relative validity.

All analytical methods used for following a process and for use as part of clinical testing should also be evaluated in terms of the theoretical basis of use, accessibility and practical considerations in addition to the ten standard indices. Running of a laboratory, in a routine mode means the four most important features limiting analysis are the customary running cost, time expenditure and staff and equipment availability. As a result of both cost and time considerations and the appropriate use of resources the four most important features of analysis are usually the correctness of the sample, the method used, interpretation of results and data storage [Hoyle (2006); Kolarik (1995); MHRA (2002); Poe (2003); Webster *et al.* (2005)].

8.3 LIMS and automation

The laboratory information management system (LIMS) represents a more frequently-used way of managing the modern-day analytical and clinical labo-ratory. A LIMS software pack is a database tailored to the analytical laboratory [Neville (2005); Wagner (2006); McDowall (1999); Friedli *et al.* (1998)] that contains integrated sourcing of batch components, sampling information and analysis in the form of a programme which generates a final report. There are a number of essential requirements of such a system, that include: secured login access to the software (limited accessibility), flexibility to add-ons and software upgrades, and perhaps most importantly, data management. This management should include, in a similar manner to conventional forms of lab management, a definition of a hierarchy of responsibility, planning, workload status, and prompting for optimised productivity, regulatory compliance and possibilities for an audit trail, reports (with appropriate calculations) and an indication of the requirements of re-testing. Most successful forms of LIMS also possess responsibility (and priority) scheduling that defines which personnel do what, where and when. In a similar manner to a 'real' lab manager the software also provides additional prompting for training, calibration, certification and material orders. The development cost of such extensive software is not cheap and the initial capital investment has to be balanced against notions of investment and impact it might have on sales and productivity. Commercial off-the-shelf LIMS software has been reported to take up to three years to be fully integrated within routine operations [Wagner (2006)] for industrial-scale production monitoring and QC.

LIMS systems are used largely because of their ability to more routinely integrate automation (and the increasing use of automation) and data handling, provide uniform methodology with complete visibility, and lead to increased productivity and process integrity that is essential in a highly regulated environment. A QA system developer should always consider the packages that are available, and user-friendly (programming) systems. Examples of the use of current LIMS software programs include, Autoscribe™ (routine QC), ClinaxysII™ (for clinical trials), Biotracker™ (biotechnology product screening using miniature 'lab-on-chip' systems), Toxchek™ (pre-clinical LIMS) and Pharmtracker™ for use in production and QC applications. It is essential that software incorporate GXP (e.g. US FDA CFR21 part 11) compliance and in this respect Pharmtracker™ has ICH compliance testing components that have a built in QA/QC modular set-up. Better quality software also provides an interface for remote operation and data transfer between sites in line with global pharmaceutical manufacturing.

It is not possible to discuss LIMS systems without discussing automation, as the two are frequently hyphenated for use in modern laboratories. Automation describes an instrumental system that involves the mechanisation of discrete processes [Buncher and Tsay (1994); Kellner et al. (1998); Selkirk (1998)] and is 'non-interventional' or self-regulating and self-timing. Many automated techniques make use of 'robotised' units and these are commonly seen in a QC laboratory for simple procedures that are susceptible to gross methodological errors, such as in liquid handling. Robotics describes the use of instrument management systems and the way in which information is handled. The initialisation of robotics use in pharmaceutical manufacture and more specifically pharmaceutical analysis started in the late 1950s where they were used to automate simple tasks in the form of continuous flow techniques to improve on lab wastage and expense. A constant stream of more recent revisions has produced automated robotised systems capable of handling multiple-channel information sources and simultaneously running selected apparatus.

Automated systems (including robotics-based equipment) have had a great uptake in routine QC and GMP practices during manufacture because of the decrease in workload (with an increase in administrative burden), decreased costs, improved institutional savings, reduced interpretive errors, increased processing capacity, consistent high levels of precision, lack of personal 'exposure' to hazardous aspects of a process and a reduction in process disruption and servicing down-time. The qualities of a generic automated system should include a minimal number of moving parts, good reliability, good equipment life expectancies, possibilities for upgrades and customer support that comply with the workplace environment, and high sample-handling capability to liberate personnel, all achieved at a relatively low cost.

Examples of widely-used robotic systems include Zymate™, Labmate™ and Benchmate™ systems for the analytical laboratory that consist of variable modules (e.g. filtration, dissolution, weighing elements and so on) that can be added or removed to assist the analyst in a series of steps within an analytical procedure. Automated systems may exist as multipurpose workstations, discrete analysers and, more recently favoured, automated miniaturised systems such as fluorescence spectrophotometric culture plate readers (DynaPro), micro-total analytical systems (lab-on-a-chip) and nano-sensors. Nano-sensors are often based on encapsulated remote devices for insertion into cells [Chattopadhay et al. (2006)] with the data handling performed outside the body. With the increasing use of remote sensing in analytical sciences this is a trend that is likely to grow considerably in future years for a number of clinical testing practices.

8.3.1 Laboratory management

Management within a laboratory is now mainly centred around use of validated software such as laboratory manager programs and LIMS-led quality control [Neville (2005)]. These give the assurance of the correct scrutiny of a product and establish a control of incoming and outgoing materials and also serve to track the status of a product during a particular process. The role of the head of the laboratory still exists but routine functions have been removed to permit personnel to dedicate more time to assessment of the results of qualifications and testing.

8.3.2 Scope and future directions

It is clear that routine QC has evolved greatly over the last ten years. The equipment used in a QC laboratory and also part of clinical diagnoses has improved, with lower levels of confident detection. A wealth of new technologies such as '-omics technology' (lipid, gene, proteome, metabol-) [Allen (2002); Yonzon et al. (2005); Balbach and Korn (2004); Greenwood and Greenaway (2002); Fujii (2002); Naylor (2004); Weckwerth and Morgenthal (2005)] now form the basis of sample examinations. Drug products and new chemical entities are evaluated more by so-called high throughput screening (HTS), analytical devices and the use of portable devices that rely on miniaturisation [Yonzon *et al.* (2005)] that might lend themselves to biotechnology-based products and PAT. The increasing use of 'smart' software (e.g. fuzzy logic) and chemometric systems including LIMS, aligned with more automated methods of analysis, provide the means to better and swifter data handling. Despite these advances there is a need for tighter regulation and the establishment of more efficient sophisticated quality systems, harmonisation and corporate integration [Sarker (2004)]. In most cases validation of the correct working of newer technologies is a principal concern before uptake and their regular routine use.

SECTION C

Starting from Scratch

9
Applications of QA to new medicinal products and new chemical entities formulation

The reason for extending a purpose-built QA practice to a new product is defined by the cost of quality or rather a lack of it. Inexperienced industrialists, pharmacy or pharmaceutical sciences students and technical staff often wonder why pharmaceuticals are so costly. There is little idea, particularly among laypersons, of the extent of testing and research involved in the drug business, encompassing development, pre-clinical, clinical testing, routine production, documentation, software, quality control, stability, formulation and excipient suitability and challenge testing steps (Figure 8.1b). All of these steps are needed, for example, for a pack of tablets or a vial full of transparent liquid that the customer makes use of. However, what needs to be stressed to newcomers to the 'area' is the compounded cost of an assurance that the product is pure, consistent and of the highest quality (PCQ) for every batch of medicine that is released from every site where the medicine is produced. In some cases this is exceptionally difficult to do and infers considerable process-related costs, nevertheless it has to be done.

Laboratory testing and all clinical testing is supported by two quality parameters, good laboratory practice (GLP) and good clinical practice (GCP). In reality 'good practice' means specific details on undertaking a process and its reporting must be made. Covering North America, Europe and all the way to Japan, the international conference on harmonisation (ICH) hopes to address inconsistencies in testing and guidelines for product PCQ [Webster et al. (1995)]. Regulators such as the FDA have long taken an interest in what might be termed routine or underpinning testing, because this is the cornerstone of all conclusions and decisions made about process and product suitability. In clinical trials an assurance is provided for a minimum of risk, and good laboratory practice forms the basis of regulatory decision making when only a limited sampling of a diverse and inherently heterogeneous group of people has been made.

9.1 Start-up and initialisation

The commencement of a project starts with the assembly of drug candidates (lead compounds); a careful selection process that leads to the best team assembly, establishment of assessment criteria and key quality concerns. This is all done with regard to addressing the overriding product (or likely product) regulations, testing and making of product-related SOPs, and adherence to regulatory compliance stipulations that is best achieved by a validation programme. Proper validation of a process is such a fundamental part of making a satisfactory end product that this cannot be stressed too strongly.

9.2 Raw materials control

The physical starting point for the preparation of new medicines is defined by the raw materials. Issues of PCQ, particularly of concern with regard to certain biotechnology products ('Biopharmaceuticals', Section 5.2.1) are critical to end-product suitability, disregarding whatever else happens to the product. Suitable materials should be based on carefully inspected (for purity and impurities), consistent raw materials and obtained from validated suppliers with a certificate of analysis (CoA) for the product. Once on site, appropriate QC testing is required, such as material identification by NIR linked to principal component analysis (PCA), or other identification systems such as bar-coding, to avoid mix-ups and aid material tracking.

9.3 The validation life cycle

The validation life cycle for each aspect of an operational module follows the process of manufacture from its conception (URS) through design tests (DQ) to mock production runs (PQ) with the aid of a strategy (VMP) and conclusions on the test or its approval (VMR). Each VMP is part of a comprehensive list of strategic goals or action points for the organisation (site validation plan; SVP). These aspects of test evolution and extensive validation across a process have been discussed in much greater detail in Chapter 4 and Section 8.2. In general, validation of a process is needed to provide the foundation for confidence in the compatibility and coherence of all the individual stages in a process of manufacture or clinical testing.

9.4 Top-down or bottom-up validations

The wisdom of validation, which at times is both painstaking and laborious but also costly, is in providing a 'guarantee' of consistency based on a form of statistical assessment and mapping of event frequency. Two general approaches are taken: for established products that are reworked to provide a newer form of essentially the same product it is possible to get away with the less desirable top-down approach to validation. This approach is also used where validation of a process or product was not possible, or insufficiently well understood in its complexity or impact at the exact time of initial examination. This is referred to as retrospective, re-validation for existing product and requires appropriate change-control documentation. The second format preferred by most practitioners is one of bottom-up or prospective validation. This format is undertaken when there is poorly known product behaviour or significant multiplicative or additive risks involved in the manufacture of a complex medicinal product [Slater (1999)]. The latter form of validation is usually preferred as a picture of product compliance or non-compliance is built-up from the simplest elements of the process to the end product. Generically, this is related to a product bearing less risk of both failures necessitating cessation of processing, and wasteful expenditure. The choice of validation method is crucial and an illustration of these two types of validation is presented simplistically in Figure 4.2, Chapter 4.

10
New products manufacturing

Preparedness for launch of a new product is the most crucial aspect, which comes from solid product or process validation and appropriate clinical trial data. Most of the regulators find themselves having to operate within local and international guidelines; in the UK this would be through the Food Act 1968 and its amendments and in North America this is through the Federal Food, Drug and Cosmetic Act 1938. In continental Europe each member state has its own reference laws (Acts, EU Directives), regulations and guidelines but potential manufacturers must also address Euro-zone legislature, coming from Brussels. Increasingly regulators operate in the frontline of the war against fraudulent manufacture, adulteration, and counterfeiting.

A number of landmark lawsuits and significant test cases have helped to sculpt both the behaviour and practice of regulators, development scientists, clinicians and manufacturers. Pivotal to this practice, were cases such as the Kefauver proposals and Kefauver-Harris amendments in the 1960s, concerning the labelling of drugs, their efficacy and safety test data. This was followed by a landmark Kelsey FDA review, which prevented the drug thalidomide from having such 'catastrophic' consequences in the US as were seen in Europe. Controversial test cases continue to be observed in Europe and the US [Schacter (2006); Rang (2006)]; in part they add to the guiding of potential future drug manufacture. Drugs featuring in the UK newspapers and media at the moment include Paroxetine (anti-depressive) and Glivec (chemotherapy) but one has to say this is surely healthy in some respects because the customer is the most important component of this complex business equation. Such sometimes-excessive negative publicity also serves to stop drug manufacturers from becoming too complacent.

The financial investment by pharmaceutical behemoths is one practice which is a delicate balance of producing a 'true good', and one in which business

Table 10.1   Some global regulators of medicines and pharmaceuticals

Area of jurisdiction	Body	Reference point (2007)
UK	MHRA (formerly MCA)	www.mhra.gov.uk
North America	FDA	www.fda.gov
France	Medicines Agency (AdM)	agmed.sante.gouv.fr
Sweden	Med. Products Agency	www.mpa.se
Ireland	Irish Med. Board	www.imb.ie
Germany	Fed. Institute for Drug & Medicinal Development (BfArM)	www.bfarm.de
New Zealand	NZ Medicines and Med. Devices Safety Authority	www.medsafe.govt.nz
Japan	Ministry of Health & Welfare	www.mhw.go.jp

viability and profit are not insignificant. Good ethics must also be foremost in high quality clinical testing; this means societal benefit (Table 10.1), lack of coercion or an imposed constraint, lack of harm, risk or injury and an overarching moral imperative from the scientists concerned. Yet the true picture is not quite as clear or simplistic as this. There are many interested parties all demanding a 'piece of the action'. Continuing development of a likely candidate drug is driven by the cost, time, associated resource investment, proof-of-principle and drug efficacy, commercialisation, mass manufacture, cost-price index, royalties, sales, competition and relative market share. A near guarantee of return on investment is needed before any organisation would gamble on producing an expensive new medicine. A patent licence would only be granted for a truly innovative medicine [Wiley-VCH and FDA (2006)]. The US Patent and Trademark Office (USPTO) and equivalents in other countries are inundated with intellectual property rights (patents; IPR) applications each year. This legal framework assures a 'monopoly of limited duration', with which the organisation can recover 'out-of-pocket' expenses. Thus, patent protection is a crucial feature of the decision to go ahead and develop new products. Extensions are granted in certain instances such a true innovation and treatment of rare disorders, in recognition of special investment by the organisation.

Production and sourcing of sufficient quantities of drug at a competitive price is a consideration that is crucial to development of an appropriate manufacturing process. Industrial-scale manufacture takes the several kilograms of pure drug used in pre-clinical and clinical trials and aims to produce a comparable formulation when the scale can be many hundred-fold larger. As such regulators like the FDA or MHRA take a keen interest in manufacture of medicines on a routine basis, and how closely these products match the product for which an approval was granted (product licence). Consequently, an operational quality system that incorporates cGMP and routine GMP standards is crucial to producing a quality product. Particular standards need to be in place for all equipment, the manufacturing environment, core testing procedures and approval. Each of these operational activities needs extensive 'audit worthy' documentation and the appropriate validation.

10.1 From inception to market place

A successful product is, in a commercial sense, one that passes from inception to the market place, passing through validation, in the most uneventful manner, with the minimum of regulation query and developmental scale-up concerns. One of the biggest growth areas for pharmaceutical development is in the arena of biotechnological use and products, hereafter referred to as biotech (biopharmaceutical products). It has been estimated that up to 25 per cent of all newly launched medicines in 2002 were biotechnology products and this is expected to exceed 50 per cent (at current growth rates) by the next decade. IPR and patenting is truly 'king' within this growth area as organisations strive to grab a market share and business foothold. However IPR provision is extremely uncertain and can be influenced by development costs or the costs of modification and subsequent re-filing.

According to recent general surveys (2004–2005) the top five organisations submitting the highest number of global patent applications in this biotechnology field were GlaxoSmithKline (GSK), Biowindow, The Department of Health USA (US DHHS), the University of California and Incite. The actual order and prevalence of organisations might have changed at the time of going to press but notably the list shows a range of commercial organisations, academic institutions and governmental departments. In the late 1990s there were 166 biotech firms in the UK that included large corporate institutions, small to medium enterprises (SMEs), and 'spin-out' organisations, such as Celltech, GSK, Oxford Molecular, Amersham, British Biotech, Corton, Porton International and Xenova.

Two 'giant' pharmaceutical organisations help to demonstrate the extent of financial and societal investment in pharmaceutical research; for convenience and because of personal dealings with these companies they are used solely as 'good company' references [FDA and GSK (2006); MHRA and Roche (2006)]. The first organisation, GSK, has approximately 100 000 employees (16 per cent deployed in R&D) with an annual R&D budget (2003) of £2.4 billion, and had 2003 sales of £21.4 billion and a pre-tax profit of £6.7 billion. The company's principal affiliations are with other pharmaceutical manufacturers such as Roche, Bayer, Powderject and Unigene that have key strengths in the therapeutic areas of anti-viral, OTC and cardiovascular medicines. The second highlighted organisation, Roche (which deploys $c.20$ per cent of its workforce in R&D) had a 2003 R&D budget of £2.1 billion with sales in 2003 of £13.6 billion and an operating profit of £2.4 billion. In this case the organisation's main business links are with Genentech, Bristol-Myers Squibb and Chugai (Swiss chemical company), based on fundamental strengths in oncology, CNS and medical diagnostics technologies.

Large organisations of this type have often gone through a number of immense corporate revisions (at a cost) and in many cases a series of acquisitions or mergers [FDA and GSK (2006); MHRA and Roche (2006)]. Previous significant mergers and changes (divestments) have included: in 1991 the SmithKline and French merger with Beecham to become SmithKlineBeecham, in 1993 ICI became Zeneca (spin-off), and in 1996 Ciba-Geigy merged with Sandoz to become Novartis. Other large-scale changes have included the 1999 merger of L'Oreal with the Total-Fina-Elf group to become Sanofi-Synthelabo (also recently involved in another merger); in 1999 Hoechst merged with Rhone-Poulenc to become Aventis; in 2000 Monsanto merged with Pharmacia and Upjohn to become Pharmacia (under an acquisition); recently BASF (Knoll) became Abbott (under an acquisition) and in 2001 DuPont became Bristol-Myers Squibb (under an acquisition) [Salvage (2002)]. All these mergers and acquisitions become significant as the diversity (usefulness) of drug business trading comes under close examination as a result of the creation of the 'behemoth' in pharmaceutical manufacture and the un-stated risk of reduced customer provision. Examples include the much talked-about landmark fusion in 2000 of Glaxo Wellcome and SmithKline Beecham to form GSK, which held 8 per cent of global pharmaceutical market. This was followed in 2003 by the merger of Pfizer (8 per cent market share) and Pharmacia to give an organisation with 11 per cent of the global market (at that point the leading and biggest organisation) and with £25 billion of revenue. At present continuing mergers are underway leading to a reduction in the number of organisations that hold the key market share and this can feel rather like a 'monopolisation.' Other globally important pharmaceutical organisations include Eli Lilly, Merck, Novartis, Johnson and Johnson, Abbott Laboratories, Bayer and Bristol-Myers Squibb [Cambridge Healthcare and Biotech (2005); European Parliament (2005); Advanstar Communications (2005); European Patent Office (2008); RPSGB (2008)]. In the middle of the decade the two top business 'players' cornered about 19 per cent of the pharmaceutical ('pharma') market. Merger and organisational growth is a significant factor, since in 2005 approximately 42 per cent (£95 billion) of the market covered the US and Canada (in the main the US), 25 per cent related to the EU-constituent states (mainly western Euro-zone states) and 11 per cent to Japan. Almost all the other regions of the world account for a mere 22 per cent of medicine consumption according to a recent survey. In this way, much financial investment is put into the diseases that affect the greatest share of the market, for obvious reasons, but this is not without justifiable ethical concerns and this would form the basis of a book in itself.

10.1.1 Requirements for a successful new product

The starting point for any new medicine involves filing for a new drug. The new drug application (NDA), which arises from successful, clear and well-represented clinical and pre-clinical data, is a document that specifies the net benefits of a NEW drug chemical entity (NCE). The NDA is a large and 'unwieldy document' taking as long as two years to assemble by a large multidisciplinary team [Amir-Aslani and Negassi (2006)]. The team can include engineers, medics, statisticians, pharmaco-toxicologists, chemists and biologists, business analysts, marketing consultants, regulators, a QA team and manufacturing experts. Clearly defined specifics are required for the active pharmaceutical ingredient (API) purity, identity, strength and PCQ of the drug product. Human studies provide crucial data on the API (active) suitability and related side-effects and risks that might limit usage.

After filing a NDA there is a long wait for approval; this usually takes from months to about a year. The decisions made by the regulatory agency concerned are far reaching as the 'weight of evidence' and 'case for support' of a new medicine are evaluated from the perspective of the participants and of a general societal benefit. The regulatory agency review panel consists of a multidisciplinary team normally comprised of an assorted array of relevant scientists, clinicians, statisticians and project co-ordinators, and the preliminary decision is customarily provided within two months of the NDA being filed. The regulators come in for much criticism from trade professionals because of the bureaucratic nature of their work. However, they do an essential job because they act as the safeguard for the customer. Their goal is to ascertain a clear picture of efficacy and justification for a new drug (not jumping on the scientific band-wagon) with a feasibility study, scientific integrity, lack of key omissions in the report and logic to the filed submission. A filing itself is not cheap and in combination the application itself, establishment and product fees can amount to $1 million (£500k). Of these by far the most expensive is the application fee, which may account for as much as 60 per cent of the fees.

A new drug product successfully enters the market provided research and clinical evaluation are satisfactory (Figure 10.1a). The product licence is only given after appropriate in-depth regulatory screening. Occasionally the products may be evaluated with a different licence application body depending on the origin of the drug. The process however, does not stop with drug approval (Figure 10.1b) for each new drug because adverse reactions and additional supplementary findings must be reported immediately. In this way the regulatory agency can continue to accrue extra support or disapproval data. All manufacturers readily participate in these studies, as they are contractually and legally obliged to monitor product safety. Product placement takes into account the form supplied

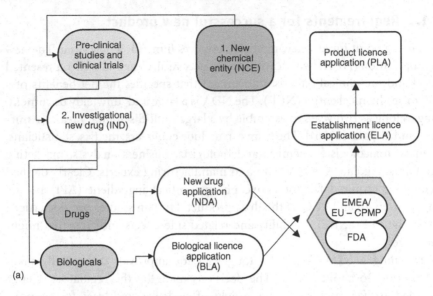

(a)

EU – EC directive 65/65EC, 1965

• Requires each member state to assess and authorise ALL medicines
• Governing organisation – European Licensing Agency (ELA)

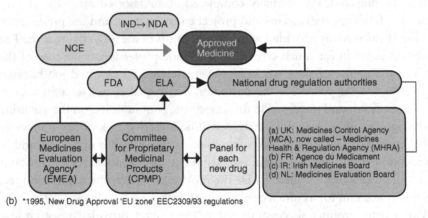

(b) *1995, New Drug Approval 'EU zone' EEC2309/93 regulations

Figure 10.1 Consisting of part (a) regulatory control in passing from concept to pharmaceutical product and part (b) licensing of new medicinal products in the US, EU and UK

to the consumer, cost and price, as related to the need for the drug, the volume of sales, product diversity and where will the drug product be targeted. The drug may be targeted specifically to locations such as clinics, hospitals or small or large commercial and distribution outlets.

An essential part of any new product development is finding the right team members. Here, a suitable mix of unconventional types of thinkers (e.g. scientists) in terms of alternative (and diverse) experiences, innovation and enthusiasm, matched against those persons having a creative approach, provides an ideal formula. Highly sought-after personal character traits in such a team include group sensitivity, analytic methodology, intuitive, observant and persevering approaches to problems. One can never underestimate the benefits to the progression of group project work where there is an atmosphere of a positive (expressive) universal involvement and both the provision of co-worker incentives and striving for continual improvement as suggested in the quality gurus' 'quality' postulates.

10.1.2 The product life cycle

The product life cycle is in form analogous to a bacterial growth curve but is formulated to show sales versus time. The relationship between product sales and lifetime always passes through a lag, growth, maturity (market saturation) and decline phase. The maturity phase represents the greatest turnover of the product and the manufacturer aims for the product to exist in this phase for the longest time. Some recycling and continuation by reformulation (Cox cycle) can keep the maturity phase going, such as conversion of tablets into topical gels, such as ibuprofen. The aim of most pharmaceutical companies is to reduce the time taken to get the compound to the market, to reduce the time for 'non-working' patent submission but position the submission early enough to maximise yield and recoup expenditure. Pharmaceutical companies are in the business of making money and providing a product of societal value but this is not always an easy balance to make. An awareness of pharmaco-economics is therefore essential to the sustainability of any organisation and its future research.

10.1.3 Innovations, patenting and intellectual property

Patenting, also referred to as intellectual property, covers three types of legal definition and intervention [Crespi (1999); Tambuyzer (2002)]. These cover elements that are:

- For a new product that can be defined and is therefore afforded full legal ownership protection
- For a product-by-process where the process is derived and can be considered new but is not easily defined with an unknown constitution
- For a process that pertains to a distinct procedure of preparing a product.

Any infringement or duplication of IPR is considered an unauthorised act pertaining to both the inventor and the good of society. The four overarching considerations in determining the legitimacy of a 'generic' patent filing are the quality, time, cost and the achieved 'value' that may be added to the products; but

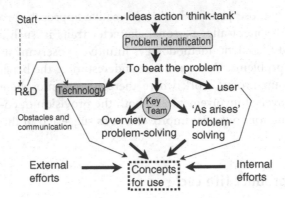

Figure 10.2 Generation of concepts using a suitable team to create usable concepts

for pharmaceutical products one might also add safety and therapeutic advantage to this list of requirements. When new products are adopted there is also a customary lag which is referred to as the 'diffusion of innovation period', which like the sale of a product with time, follows an exponential growth as increasing adoption takes place but might also end in societal saturation of the technology.

Appropriate resolution of a therapeutic drug delivery system shortage or shortfall might be solved by creating an appropriate think-tank of 'real experts' that allows the fullest organisational networking (Figure 10.2) and technology transfer from cutting-edge R&D findings to technology application. This has certainly been the case with the array of stealth liposome medicines (patents, current developments) and information fusion with material and biochemical sciences. Problem solving is often best achieved by formulaic combinatorial assembly of problem recognition and an appropriate project team, but this has to be pitched against the costs of development. In 2006 a patent realisation review showed it took typically 10–12 years to develop ONE new drug at a UK cost of £0.4–0.5 billion. Therefore, in real terms a company must be able to secure marginally greater than $1 billion (£530 million, 780 million Euros) to fund the process with a degree of security. In this case about eight years remains (maximum) of proprietary patent for the manufacturer to recover the associated expenditure. Development of new drugs costs so much because, for example, in the US (in 2006) each clinical trial subject in the past was 'costed' at approximately $10k (£5.4k). The number of patients has to be considerable in clinical screening to provide the organisation with a reasonable set of statistically valid (valid to the patient population) results.

Drug development is an extremely inefficient (and costly) process that can best be illustrated by the following approximated model, since only about 0.01 per cent candidate (lead) molecules reach commercialisation. In a generalised screening scenario where there were: (a) 10 000 compounds synthesised, of these only (b) 15 would meet key criteria, of which (c) 6 would meet the chemical and biochemical demands of pre-clinical development, that gives (d) a mere 3 that would be effective in human models, leading to (e) only 1 remaining candidate drug after comparison with current products. All products must be reviewed in terms of their side-effects, synthesis cost and pharmacological efficacy in addition to product surveillance and market tests. As such the cost of new products carries about 70 per cent of the costs attributable to failing entities within the organisation. In addition employment of scientists, clinicians and regulators within an R&D context helps costs to spiral. In the UK in 2006 approximately 25 per cent of all R&D expenditure was pharmaceutically related with financial support coming in different forms from the UK research councils, indirectly from government and directly from industry. The UK only spent 2.2 per cent of its gross national product on technology research in 1992 and most of this did not pertain to pharmaceutical development. In a global context in 2005 private organisations spent a mere £35 billion on pharmaceutical research and development and of this 40 per cent went on clinical trial expenditure.

Technology transfer provides a means of circumventing huge developmental costs. Such schemes make increasing use of joint venture programmes, where private organisations, the public sector and university consultancy groups operate through academia–industry collaborations (AICs). These collaborations are used increasingly because of poor local research funding, the opportunities for IPR and business opportunities. The business opportunities for a company allow an external pool of expertise to be used to get around large numbers of potential lead candidate molecules, process inefficiencies and spiralling (set-up and validation) costs. Each side of the collaboration has key considerations; academic bodies prioritise in terms of technology, reputation and publishing and companies consider technology, commercial risk, IPR and market intelligence to be the most important, in general.

Research projects are conventionally configured according to identification and disease targeting, culminating in biological testing that is based around the R&D process (Figure 10.3). The assembly of initial study information, a project team, a non-wavering project steering group (to keep the project within timelines) and a commercialisation group, is used to get to the point of having a pharmaceutical NCE.

The potential drug developer's considerations include an assessment of the organisation's possible advantage (opportunities) or risk of development, which is a major part of the product being novel, the 'driving through' of the project, product timing (existing product life cycle) and the quality of concept (new model concept matching the needs against the technology available). However, the cost and value of the 'return on investment' to the partners in an association or the organisation often come out on top in terms of individual feature weighting for obvious reasons. The attributes and features are put together in a target product

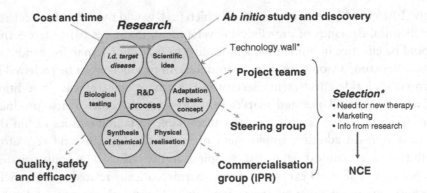

Figure 10.3 Phenomenological model of research and development processes relevant to new pharmaceutical product development

profile (TPP) or product innovation charter (PIC). There is always a balancing of product opportunity against the resources that are at hand, set against the risk and cost of an innovative concept and the extension to a new product. In a highly simplistic evaluation of the key criteria organisations appear to primarily seek significant market share and royalties, profit, sales (that relates to net income) and an 'innovation stream' to out-perform competitors (see competitive advantage as described in Figure 2.2).

New products relate to both intellectual property and development and clear proof-of-principle for a new concept [Rang (2006); Gibson (2004)]. They exist in two formats: 'true' innovation (EU patents are generally 20 years) where 'products' must be both novel and applicable, and a 'product' re-working (reformulation). Much current pharmaceutical research makes use of contract research organisations (CROs). This is driven by the extreme cost of phase III in clinical trials and the return on investment. Such outsourcing has the disadvantages of lack of accrued experience and commitment, occasional loss of effective dialogue or communication and provides a 'short-term only' gain contrasted against the advantages of reduced overheads, use of 'true' experts, process optimisation, reduced lead (development) time and the liberation of personnel and resources.

10.2 New product development: product design and specification

Drug development takes place in four extensive and mutually exclusive stages referred to as phases I (one) to IV (four). Phase III involves the appropriate activities pertaining to commercial scale-up, and phase IV is generally accepted to encompass process validation and product launch. The initialisation of the process, called phase 00 (Figure 8.1b) involves the lead compound disclosure and

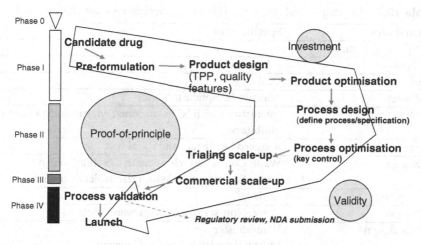

Figure 10.4 The pre-clinical, clinical and industrial validation aspects (phase 0 to phase IV) of a pharmaceutical new product development, and negotiation of obstacles prior to launch of a new medicinal product. Adapted from [Gibson (2004), 157–173].

discovery. The basic process of NPD involves 5 steps that are: the *opportunity* (market survey), *concept generation* (selection based on opportunities), *concept and project evaluation* (marketing, financial, technical aspects), *development* (technical product needs, marketing) and *launch* (commercialisation). The milestone steps for passage of candidate drug to product launch are shown in Figure 10.4.

A product design report (PDR) is made after product development and it gives details of patents and IPR, quality details and design specifications, any risk analysis of technical aspects and investment, safety information (concerning toxicology), environmental health issues and either a target product profile (TPP) or minimum product profile (MPP). These form the skeletal basis of *licence to tender* and produce a new entity in terms of regulatory compliance, any pre-approval inspection (PAI) and the ultimate legally-binding document that is the *product licence*.

10.2.1 The target product profile

The TPP is a summary definition of product attributes, customer and end-user needs that is based on customer 'wants' and 'needs' (refer to Maslow's (1943) hierarchy of needs) fitted against the capabilities of the manufacturer [Crawford and Di Benedetto (2006); Reinertsen (1997); Kennedy (1997); Drews and Ryser (1997); Ottosson (2004)]. It makes use of strategic planning to link the most desirable attributes of a product to a strategy for new products (product innovation charter, PIC) via goals and objectives. The PIC routinely

Table 10.2   The target product profile (TPP) or product innovation charter (PIC)

Attributes	Specification
Disease/indications	'x'
Client	Adults age 'x'
Administration route	Oral
Efficacy	Selective 'x' uptake inhibitor
Safety	Interactions with 'x', 'x' enzymes, enzyme inducers and inhibitors
Economics	Reduced healthcare and social cost
Dosage	Want: Controlled release matrix, 'x' film-coated Aesthetics: colour code with size/tablet strength
Frequency	'x' daily
Process	Want: Non-standard compression (details)
Pack design	Want: Blisters Must have: with barrier, self-opened
Distribution	US, EU, Japan
Costings	Want: goods not more than 'x' % commercial price Must have: price equivalent of 'gold standard'

contains relevant background (key ideas), focus, goals and objectives and project guidelines, and only has the additional feature over a TPP of specifying some details of possible manufacturing procedures (Table 10.2).

An appropriate TPP for any new product development requires a key development team, appropriate proportioning of creative and critical inputs that are required for best outcome. In most cases relevant to new drug development and the TPP the project directorate should have full representation of the following: chemical scientists, drug kinetics, metabolism and toxicology experts, formulation and development expertise, pharmaco-economics and marketing consultation, medics (clinicians and nursing staff) and regulatory affairs guidance.

10.2.2 Quality function deployment

Quality Function Deployment (QFD) is used for product engineering and came (like many business practice approaches) from Japan in the 1960s and then to the UK in the 1980s. QFD provides a strategised means of quality improvement by three elements: a focus on customer requirements (subjective), a focus on company configuration (quantitative) and a focus on design characteristics (specifications). The approach is often referred to as the 'house of cards' chart, mainly in respect of is appearance (Figure 10.5). Other considerations of QFD encompass notions of a product risk analysis, finer aspects of product safety and the proposed shelf life of the commodity.

QFD utility lies in its ability to permit a matriculated form of production planning that uses a structured approach to risk assessment, a tabulated form

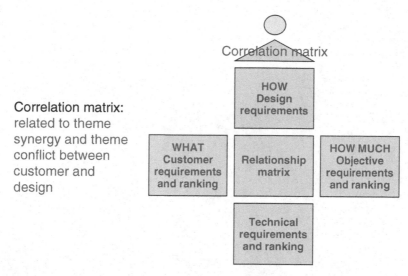

Figure 10.5 Meeting the product, customer and manufacturer's needs through a matriculated assessment referred to as Quality Function Deployment (QFD)

that considers unit operations, focus, goals, guidelines, background and ranking, and finally a multi-faceted 'gap' analysis of weaker points in the process. Venture capital development examples (accounting for about 5–10 per cent), which currently use and continue to use a QFD approach, include nano-medicinal drug delivery systems and cosmetics. As a consequence of safety concerns, 'nanotechnology' image association and regulatory concerns this approach can aid assessment of the feasibility of a product even before getting inside a laboratory [Gewin (2006)].

Unsuccessful projects and NPD are usually based on five failings and these are: a flawed underpinning science base, poor planning, safety concerns (in the biomedical sciences this is the number one cause of failure), poor marketing intelligence and customer focus, and poor project control and management (including financial management). Needless to say better and less wasteful approaches to development amalgamate proven assessment strategies such as QFD and TPP with very careful assessment of the weaknesses and strengths associated with both the organisation and the potential new product. None of this works well without the appropriate composition of the development teams and management of the scientific and personnel-related aspects of projects, and ultimately this can impact on the data suitability and the quality of the new products and new product evaluation.

Figure 20.5 ...

11
Questions and problems

These are provided as an aid to the student or newcomer to the areas of industrial pharmaceutical practice.

11.1 Specimen examples and exam questions

A selection of real and potential assessment material for the reader that should aid understanding of the field, integration of subject themes and add to material contained within chapters. Information to answer these questions is contained in relevant sections within the book.

11.1.1 Section A – coursework

Level '3' BSc Pharmaceutical and Chemical Sciences – course assignment

The answer MUST follow explicitly the guidelines given:

Guidance notes:
Please read through these guidance notes to the exercise below. Write a 1000-word report in an essay format on the theme detailed below.

Your work must be typed, and must have the following general sections: introduction, discussion and summary. You should also include a word count, schematics where appropriate and must include appropriate referencing. Group activity and plagiarism are *not* acceptable in this work.

Your work will be marked with respect to originality (lack of plagiarism), technical merit, critical and comparative discussion, drawing of appropriate conclusions, provision of worked examples and evidence of background research and reading.

Copying/evidence of plagiarism and excessively low/high word count (>5 per cent lower/higher than limit) will be marked down.

Background:
Schematic showing a surfactant/lipid-based micelle, the unit part of a colloidal drug delivery system with polymer (biopharmaceutical) encapsulated in the centre. This should be considered as part of your answer.

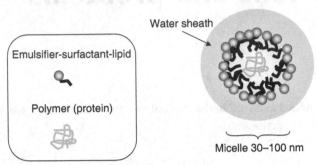

Emulsifier-surfactant-lipid

Polymer (protein)

Water sheath

Micelle 30–100 nm

There are greater than 10^6 micelles per sample

Micellar Encapsulation

Title:
A pharmaceutical manufacturer is routinely producing a sterile biopharmaceutical product. The nature of the product is an aqueous, buffered, ampoule-based form of an encapsulated protein drug. Describe the concerns of routine effective Good Manufacturing Practice (GMP)-grade production and the remedial action, if any, a manufacturer of bulk biopharmaceuticals might take to ensure the best quality product. You may assume the manufacturer operates within a Total Quality Management (TQM) framework, adhering to standards, such as ISO9000 and ISO9001. Write a structured précis account concerning the most important issues.

Some appropriate texts in the first instance may include:
Books:

1. Kolarik (1995) *Creating Quality: Concepts, Systems, Strategies and Tools*, McGraw-Hill.

2. MCA (1997) *Rules and Guidance for Pharmaceutical Manufacturers and Distributors 1997*, MCA, Stationery Office Ltd, 3–177.

Journal articles:

3. Johnson (2003) *Pharmaceutical Technology Europe*, December, 57–61.

4. Doblhoff-Dier and Bliem (1999) *Trends in Biotechnology*, **17**, 266–270.

5. Sarker (2004) at www.touchbriefings.com/pdf/890/PT04_sarker.pdf.

6. Sarker (2005a) *Current Drug Delivery*, **2**(4), 297–310.

'M' level Pharmacy – course assignment

The answer MUST follow explicitly the guidelines given:

Guidance notes:
Please read through these guidance notes to the exercise below. Write a 1000-word report in an essay format on the theme detailed below.

Your work must be typed, and must have the following general sections: introduction, discussion and summary. You should also include a word count, schematics where appropriate and must include appropriate referencing. Group activity and plagiarism are *not* acceptable in this work.

Your work will be marked with respect to originality (lack of plagiarism), technical merit, critical and comparative discussion, drawing of appropriate conclusions, provision of worked examples and evidence of background research and reading.

Copying/evidence of plagiarism and excessively low/high word count (>5 per cent lower/higher than limit) will be marked down.

Title:
1. You are provided with two alternative procedures for improving the quality of a topical oil-in-water product:
    (a)   Improve the manufacturing procedure according to ISO 9000/9001.
    (b)   Reduce the number of defective and poorer quality (non-conforming) raw materials, by appropriate procedures, used in an unchanged manu-facturing procedure.
    (c)   Discuss, with reasons, which of these two approaches above should be adopted to improve the quality of the final product. [13 marks]

2. Discuss the basis of the 6-sigma control and quality system and relate this to routine process control for the production of a sterile colloidal drug delivery system. [7 marks]

Some appropriate texts in the first instance may include:
Books:

1. Harris (2003) *Quantitative Chemical Analysis*, WH Freeman & Co., pp 720–723; 733–739.

2. Kolarik (1995) *Creating Quality: Concepts, Systems, Strategies and Tools*, McGraw-Hill.

3. MCA (1997) *Rules and Guidance for Pharmaceutical Manufacturers and Distributors 1997*, MCA, Stationery Office Ltd.

Journal articles:

4. Johnson (2003) *Pharmaceutical Technology Europe*, December, 57–61.

5. Doblhoff-Dier and Bliem (1999) *Trends in Biotechnology*, **17**, 266–270.

6. Sarker (2004) at www.touchbriefings.com/pdf/890/PT04_sarker.pdf.

7. Sarker (2005a) *Current Drug Delivery*, **2**(4), 297–310.

11.1.2 Section B – exam MCQs

1. Describe what you understand by the term validation:

   A   Two types, conformity of process

   B   Three types, conformity of process

   C   Two types, non-conformity of process

   D   Three types, non-conformity of process

   E   Four types, non-conformity of process

2. The acronym GLP in connection with process testing represents what?

   A   Good Laboratory Protocol

   B   Governmental Legislative Practice

   C   Good Laboratory Practice

   D   Given Laboratory Protocol

   E   Greater Laboratory Practices

3. What do you understand by the industrial acronym TQMS?

   A   Validation via trends, qualities, materials and standards

   B   Total quality material standard

   C   Total quotient of manufactured samples

   D   Total quality material system

   E   Total quality management system

4. Cleaning validation acceptance criteria use the formula, $C = (dbF)/d_n$ to calculate the maximum carry-over concentration (C) permitted per 100 cm^2 of contact surface. If d, b, F and d_n represent: therapeutic dose, batch size, safety factor and largest daily dose of new product to be manufactured, respectively, then calculate C if: $d = 100$ mg/tablet, $b = 100$ kg, $F = 0.1$ and $d_n = 250$ mg/tablet. What type of product is this likely to be, based on the safety factor?

   A   4.0 mg, parenteral

    B   40 mg, parenteral

    C   25 mg, topical

    D   4.4 mg, topical

    E   4.4 mg, solid dosage form

5. GCP in connection with drug development and testing means:

    A   Good clinical practice

    B   Good characterisation of practice

    C   Given characterisation of protocol

    D   Good commercial practice

    E   Good chemical practice

6. A manufacturing SOP governs:

    A   System operating in practice

    B   Standard operating procedures

    C   Standard organisational structures

    D   Safe operating practices

    E   Suitable official protocol

7. What does the acronym ISO9000 represent?

    A   Internal safety organization, specific guidelines

    B   International safety organization, specific guidelines

    C   Internal standards organization, specific guidelines

    D   International standards organization, specific guidelines

    E   Internal screening organization, specific guidelines

8. The term batch document within a cGMP environment relates to:

    A   A document which follows all produced medicine

    B   A document which follows QC activities relating to a batch of produced medicine

    C   A document which follows a batch of produced medicine

    D   A document which scrutinises a batch of raw material

    E   A results sheet relating to the quality of manufactured goods

9. What are the three most important considerations of a validated analytical method?

    A   Precision, accuracy, specificity

    B   Precision, specificity, linearity

    C   Linearity, limit of detection, robustness

    D   Specificity, ruggedness, robustness

    E   Linearity, ruggedness, robustness

10. Quality assurance in drug development uses an expression – M2I (MII or MI^2); this represents:

    A   Manage-innovate-improve

    B   Maintain-manage-improve

    C   Maintain-innovate-invigorate

    D   Maintain-innovate-improve

    E   Modify-innovate-improve

11. The three principal established quality gurus that initiated the 'Quality Culture' are:

    A   Shewhart, Deming and Müller

    B   Deming, Juran and Crosby

    C   Jeming, Duran and Crosby

    D   Ishikawa, Feigenbaum and Jenning

    E   Ishigawa, Müller and Shewhart

12. Shewhart's cycle consists of (order is crucial):

    A   Plan-do-check-act steps

    B   Do-plan-act-check steps

    C   Plan-act-do-check steps

    D   Check-plan-act-do steps

    E   Plan-wait-do-check-act steps

13. In an analytical validation sense what do precision and accuracy, respectively, really mean?

    A   Bias, sensitivity

    B   Statistical error, systematic error

    C   Bias, statistical error

    D   Data trueness, individual measurement data variance

    E   Individual measurement data variance, data trueness

14. What do you understand by the term quality control?

    A   'Planned system of activities to provide a quality assurance'

    B   'Paradigm of activities to provide a quality assurance'

    C   'Planned system of activities to provide a quality product'

    D   'Paradigm of activities to provide a quality improvement'

    E   'Proposed activities to provide a suitable qualified person'

15. Cleaning validation and manufacturing suitability is established universally via two generic sampling methods, these are:

    A   Sample placebo limits, surface concentration limits

    B   Sample placebo limits, settle plate limits

    C   Swab method limits, sample placebo limits

    D   Sample placebo limits, surface appearance limits

    E   Air impinger limits, glove method limits

16. What do you understand by the term Kaizen?

    A   Expert panel/quality circle

    B   ISO system

    C   Audit readiness/TQMS

    D   Validated methodology/management system

    E   ISO CRM standards

17. Biopharmaceutical products include:

    A   Vaccines, hormones, *cis*-platins, cytokines, growth factors

    B   Vaccines, hormones, tissue extracts, enzymes

    C   Vaccines, hormones, *cis*-platins, cytokines

    D   Vaccines, hormones, *trans*-platins, toxoids, enzymes

    E   Vaccines, hormones, *cis*-platins, toxoids, enzymes

18. Process validation acceptance criteria use the formula $S = n(AQL)$ to calculate the suitability (S) of a process, in this case for a potent cytotoxic medicine. Here, quality is assessed as non-conformance (defectives) in terms of dosage strength nominal content. Acceptable quality level (AQL) is a *quotient* but is often expressed as a percentage for convenience. If n and AQL represent: the sample lot size and an acceptable quality level, of 1×10^6 units and 1.5 per cent, respectively, does the tested medicine pass if you measured 20 000 defectives in the lot?

    A   Neither pass nor fail, re-test

    B   Pass

    C   Clearly pass

    D   Clearly fail

    E   Not related

19. In a statistical sampling plan, known as an operating characteristic curve (OCC) the *probabilities* of accepting-unsatisfactory batches or rejecting-satisfactory batches of medicines are referred to as:

    A α – producers' error; β – consumers' error, respectively

B β – producers' error; α – consumers' error, respectively

C β – consumers' error; α – producers' error, respectively

D α – consumers' error; β – producers' error, respectively

E χ – consumers' error; ζ – producers' error, respectively

20. Typically 6-sigma processes encompass what percentage of the data, in a normal distribution function?

A   99.9999998 per cent

B   0.0000012 per cent

C   99.74 per cent

D   68.27 per cent

E   95.45 per cent

21. For parenteral medicines it is generally recommended that terminal sterilisation use what grade of microbial load starting material?

A   A

B   D

C   G

D   C

E   B

22. An alternative to the operating characteristic curve is a graphical sampling plan proposed by:

A   Vervey and Overbeek

B   Jones and Jenkins

C   Hansen

D   Langevin and Schmidt

E   Dodge and Römig

23. TPP and QFD respectively stand for:

A   Target Project Profile, Quantick-Fiddes-Dean method

B   Total Product Performance, Quality Function Deployment

C   Target Project Profile, Quality Factor Dysfunction

D   Target Product Profile, Quality Function Deployment

E   Target Project Profile, Quality Function Displacement

11.1.3 Section C – exam short and long answers

Short questions

1. What do you understand by the term QA when applied to medicine manufacturing?

2. Describe what you understand by the term 'process' validation.

3. Explain the term current Good Manufacturing Practice in connection with routine pharmaceutical production.

4. Explain the term Good Laboratory Practice in connection with process testing.

5. What do you understand by the acronym TQMS?

6. What do you understand by the term quality circle?

7. Describe what you understand by the term analytical validation; give one example.

8. Given an experimental average of 210.5 tablets per batch, a standard error of 6.1 tablets per batch and a 95 per cent confidence interval (giving a t-test value of 1.96), what would be the number of samples needed to be taken? (Where, $n = (x_{bar}t)^2/e^2$.) Of what value would this validation study be if the sampler chose to take 2000 measured samples?

9. Describe two numerical indices of sampling. You need formulae, and to describe their application and relative value.

Long questions

1. Given the following 'details' of an intensively processed new medicine and process limitations for such an investigational new drug (IND) product, discuss risk prevention and end product quality assurance with the aid of suitable examples: [20 marks]

   Raw material and process details:
   (a)   Chemically labile biopharmaceutical and excipients
   (b)   100 per cent testing needed
   (c)   >6-sigma control needed
   (d)   Heat sterilisation not possible but product must be sterile
   (e)   Product is to be filled, processed and labelled using 'new' plastic MDPE ampoules.

2. Critically discuss *two* of the following, making use of examples: [10 marks each]
   (a)   cGMP
   (b)   Production-related documentation
   (c)   PCQ
   (d)   QC charts, related to medicine production.

3. Given the following generalised process outline, discuss how a 'solid validation approach' may be used to reduce the risk of an out of process-control batch of sterile ampoules for injection.   [20 marks]

Process:

START: incoming materials → raw materials → filtration → assembly/formulation → filling → retorting → labelling → cartoning → dispatch to customer – FINISH.

4. Critically discuss the 'quality control spiral', the commonly used expressions 'total quality assurance' and 'quality is everyone's business', and the key duties and responsibilities of a Qualified Person.   [20 marks]

5. Discuss the concepts of Quality Assurance. Include in your answer examples of its relevance to patient care in a hospital.   [20 marks]

6. What pharmaceutical/chemical data are required to provide a good understanding of the characteristics of a new drug substance prior to formulation into a product? Show how these data may provide guidance in minimising formulation difficulties?   [20 marks]

7. Critically discuss *three* of the following, making use of examples: [6 marks each, 2 marks for examples]
   (a)   Statistical basis of sampling
   (b)   Duties of Qualified Person
   (c)   Validation
   (d)   Quality improvement strategies and customer feedback
   (e)   Cleaning validation
   (f)   SOPs.

8. Given the following outlined 'sketch' of a process for a new drug product (phase IV of development), what would you consider to be the issues of generic concern for a manufacturer?   [20 marks]

   Process outline:
   (a)   START – raw materials (5), including products from 'natural source' for liposomal encapsulation of drug
   (b)   + potent cytotoxic biopharmaceuticals (2)
   (c)   Assembly and filling* (ampouling) – many sub-steps
   (d)   Sterilisation
   (e)   Packaging
   (f)   FINISH – storage for distribution.

   * makes use of a clean room environment and isolator technologies.

9. Critically discuss the essential differences, merits and point of application of top-down and bottom-up validation in pharmaceutical manufacturing. Provide examples of where each might be used.   [20 marks]

10. (a)   Given the following data set, construct an operating characteristic curve:

Number of defects	Probability of accepting batch (at AQL)
0	1.000
3	0.920
6	0.850
9	0.710
12	0.470
15	0.350
18	0.310
20	0.300
24	0.295

Find the probability that a batch with 10 per cent non-conforming units will be accepted. If the lot tolerance per cent defective number is 17 per cent and the acceptable quality level is 8 per cent, what are the risks of consumers' (β) and producers' (α) errors, respectively, as percentages? [7 marks]

(b) Strictly speaking, what is an operating characteristic curve and why might it be needed? [3 marks]

(c) Describe the sequence of events in a routine process validation, highlighting any areas worthy of particular attention. Illustrate your answer with examples. [10 marks]

11. (a) Draw the control chart (QC chart, Shewhart 3-sigma format) for the following data set. You should critique your findings with regard to statistical or systematic variation and comment on process suitability and the effectiveness of the quality improvement measure. [15 marks]

Data:

Batch	Content of active drug substance (mg/tablet), based on 10 tablets
0	251.0
5	259.5
10	260.1
15	260.2
20	263.5
25	267.8
30	264.4
35	260.5
40	258.4 ← improvement to process

Batch	Content of active drug substance (mg/tablet), based on 10 tablets
45	254.3
50	252.1
55	247.9 ← improvement to process
60	245.8
65	249.6
70	249.7
75	250.5

To define the process you should use the established nominal values:

Average	250.0 mg
Standard dev.	4.2 mg

(b)   Discuss the specific role of the QP with respect to documentation and change control. What measures would a QP need to take if the process was optimised by a subtle change of process that still complies with the product licence?   [5 marks]

11.2 Model answers to examples

11.2.1 Section A – Degree and 'M' level coursework

Model answer coursework (1)

Work marked on:

1. Technical correctness

2. Analysis and 'critical' and 'comparative' discussion*

3. Clear conclusion

4. Concise and amplified (weighted) at appropriate points in relation to *specific* question themes.

*Discussed themes must include:

- Balance of Pharmaceutical RMs to process change/appropriate points clarified (*most important*)
  - Can a poor RM lead to a good end product? Example. How important is this for initial or terminal sterilisation? Significance? Problems of formulation (*most important*) for the sterile product, e.g. loss of activity of conformer.
  - Example of ISO relevant to quality improvement for such a product.

- It is important to illustrate/contextualise points with reference to examples of the product type (pharmaceutical) given in the question (*most important*).

- PCQ (*most important*) issues for this and other relevant examples, given ingredients.

- Underpinning systems – GMP (GLP, TQMS), but only in so much as is relevant to the question. This does not answer the question but adds background to a specific answer.

- Justification of conclusions.

- Briefest description of what process control (6-sigma) is, but some mention of probability of error and systematic variation (due to the nature of the biopharma product).

- Process control systems – are they relevant to a *sterile* product! Alternative remedial action (to 6-sigma driven sterilisation) should also be proposed e.g. irradiation, asepticity in production, etc.

- Overlapping lab/production tests leading to TQC.

- Examples must be used to support conclusions (*most important*).

Model answer coursework (2)

Work marked on:

1. Technical and regulatory correctness

2. Analysis, and 'critical' and 'comparative' discussion*

3. Clear conclusion

4. Concise and amplified (weighted) at appropriate points in relation to *specific* question themes.

*Discussed themes must include:

- Balance of Pharmaceutical RMs to process change/appropriate points clarified (*most important*)
  - Can a poor RM lead to a good end product? Example. Is a topical sterile? Significance? Problems of formulation (*most important*) for topical (1) and the sterile product (2).
  - Process is already operating under ISO/9000/9001 systems (GMP) or any other regulating guidelines, according to the question – this should be assumed (because of 'improving'). Example of ISO relevant to quality improvement.

- It is important to illustrate/contextualise points with reference to examples of the product type (pharmaceutical) given in the question (*most important*).

- PCQ (*most important*) issues for this and other relevant examples, given ingredients.

- Underpinning systems – GMP (GLP, TQMS), but only in so much as is relevant to the question. This does not answer the question but adds background to a specific answer.

- Absolute conclusion of preferred method (a)/(b) (*most important, 2 marks*). Justification.

    Parts (1) and (2) may be integrated (or written in sections) as there is overlap, and this should be mentioned, but part (2) must mention specifically:

- Briefest description of what 6-sigma is, but some mention of probability of error and systematic variation (sketch would be useful).

- DMAIC, c_p and relevance to a *sterile* product! Is it relevant? Not fully.

- Alternative remedial action (to 6-sigma driven sterilisation) should also be proposed, e.g. irradiation, asepticity in production, etc.

- Overlapping lab/production tests leading to TQC.

- Also relevant to part (1) in that RM purity is paramount – likely influence?

- Examples must be used to support conclusions (*most important*).

11.2.2 Section B – exam MCQs

1. B

2. C

3. E

4. A

5. A

6. B

7. D

8. C

9. A

10. D

11. B

12. A

13. E

14. C

15. A

16. A

17. B

18. D

19. C

20. A

21. B

22. E

23. D

11.2.3 Section C – exam short and long answers

Short questions

1. System of planned activities; ensures PCQ, methodology.

2. Conformity of 'qualified' process. Suitability to manufacture.

3. Regimen of production/testing/protocol activities to assure medicine quality.

4. GLP is regimen of laboratory activities to assure result quality.

5. Total Quality Management System e.g. Kaizen philosophy, integrated holistic 'QA/QI (best R&D)' methodology; uses standards, might include LIMS or other formal structures.

6. Quality circle relates to total quality management system e.g. expert panel/ Kaizen, integrated holistic 'QA/QI/R&D' methodology, uses standards to recommend best approach.

7. Analytical validation – consistency of process, fitness for purpose, infers best method that has uppermost specificity with appropriate sampling (confidence in results) e.g. method accuracy, precision, linearity, robustness, ruggedness, selectivity, etc.

8. Answer: 4575, under-represented/reduced confidence in result.

9. Answer: $n = \sqrt{}$ (total number samples), rule of thumb; $n = ((s \times t)/E)^2$ – confidence interval, correct.

Long questions

Points discussed in the correct answers to questions highlighted should include expansion on these keywords, phrases and notes:

1. Overall: raw material – PCQ; then: validation (new drug – role of QP, documentation, analytical testing) – format, supply chain, hygiene/asepticity, sterilisation without product degradation (heat not required? Co60, filtration 0.22 micron, ethylene oxide, hurdle technology, not pH!), clean room HEPA/HVAC/isolator technologies? Shelf life, packaging breach. Examples of comparable case studies required. Nature of product, specifics related to scenario!   [20 marks]

2. Answers: (a) Conformance/suitability/legality/regulation-training; (b) Auditing, SOPs, batch documents, validation plans, guides. Change control, points/testing; (c) PCQ = purity-consistency-quality at all stages (expand); (d) X/R-type charts, trend following – use. ALL – drawbacks and pluses needed. Example of advantage is consistent method, and disadvantage is bureaucracy/failing in communication.   [20 marks, 10 marks each]

3. Solid validation approach follows validation life cycle – URS, DQ to PQ. Must involve review and replication. Must involve diverse team (skills, know-how) and involve QP. HACCP and HAZOP analysis required. Zone demarcation and process linearity mandatory to prevent cross-contamination. Risk highlighted from method transfer and validation. Sterile products carry particular risk (PCQ) of RMs to final product and notion of mid-, or terminal sterilisation. Type of sterilisation may be an issue. Heating may not be possible, consider irradiation of EO basis. Packaging and freedom from risk of cross contamination an issue.   [20 marks]

4. (a) The QC spiral: a series of discrete iterative practices, adopted by a quality circle. Its mission is one of Kaizen. Continual improvement in regulation, compliance and quality mean the product ultimately achieves perfect status. Whether this is arrived at is a contentious point. (b) Total QA means QA taken as a holistic integrated system, in some respect it is pro-active and all-encompassing. It can be called proper QA. (c) Quality is everyone's business means sticking to points of quality gurus' (DJC) mission statements, and efforts to comply with right-first-time. No one person in the 'quality loop' is exempted. (d) Qualified Person duties are: seeing manufacturer licence is obeyed, batch conformance (PCQ), local expertise, validation and training, auditing, batch sign-off, SOP and batch documentation, regulatory compliance and change control. QP responsibility is to profession-organisation and CUSTOMER.   [20 marks]

5. Quality Assurance. Sampling and adequate testing fall at the forefront of activities. Involves: Includes plan-do-check-act Shewhart's cycle as part of

review and validation exercises. ISO and local standards in place. Consists basically of *validation* and *quality control* but with document control, product release, CPD and staff training; all of which are overseen by the Qualified Person/suitable manager. QA and hospital-based quality and pharmacy personnel work alongside clinicians to optimise patient care in a hospital. They must address problems associated with environment and specials manufacture. Key elements not seen in community pharmacy include: dressings, implants, diagnostic kits/imaging aids, steriles, vaccines, radiopharmaceuticals, medical gases and anaesthetics. PCQ rules apply here as throughout any production/delivery process but risk is increased so stringency should be also. [20 marks]

6. Chemical data: optimum solubility, form, purity, pH/pKa, dissolution profile, crystal form, compressibility/compression characteristics, stability (light, pH, polarity), sensitivity to key catalysts, methods of identification and analysis. Pharmaceutical data: toxicity, side-effects, dosage profile, likely route of administration (clinical trials data), compatible and incompatible ingredients and excipients, drug antagonists and synergists, pharmacology or mode/site of action and ADME are required for new drug substance prior to formulation into a product. Knowing why/how/where and when the drug product will be administered helps a formulator 'package' the drug appropriately and avoids failure or rejection in the later stages of product validation and pre-launch/launch. Knowing these will avoid poor formulation and optimisation via QFD and TPP. [20 marks]

7. Breakdown: (a) Sampling coefficients, OCC/tabular/QC charts, hypothesis tests, estimates/validity/bias test – use; (b) Conformance/suitability (PCQ)/ legality/regulation-training, documentation; (c) Conformance based on rigorous testing (SOP), DQ to PQ, Validation Plan and HACCP control points/testing, PCQ/QS ISO9000/1/2/4; (d) Quality control spiral, quality circle. Integrated R&D/QA; (e) Suitability for manufacture, various limits (clean, 1 ppm, 0.1 per cent dosage, etc.), select analytical/microbiological methods of testing, CIP/WIP issues; (f) SOP = 'exact protocol/guide', e.g. DQ to PQ, VP/VMP/VMR. [6 marks each, 2 marks for examples]

8. Overall: PCQ; then: material consistency – e.g. egg phospholipid, plant/tissue/blood extracts (safety)! Encapsulation ratio = potency. Validation (new drug poses particular problems) – format, supply chain, hygiene/asepticity, sterilisation without product *degradation* [polymorph, encapsulation ratio, solubility excipient/active]. Sterilisation is the key risk area (heat, Co60, filtration, EO?), clean room/zonation? Shelf life, packaging corruption, role of QP, docs, analytical testing. Cytotoxic drug requires training and isolation/site dedication/specialised environment. Examples of suitable case studies would

be good e.g. doxorubicin, daunorubicin (daunoxomes), paclitaxel, rhizoxin, podophylotoxin, penclomidine, virosomes.   [20 marks]

9. Validation: Top-down – retrospective; bottom-up (normal) – prospective validation. Re-validation would normally be bottom-up. Risk, margin for error, value of goods considerations. Sketch of format or clear explanation. 'Similar' products might assume top-down based on prior investigation – safety judgement suggests risk factor important for perhaps low potency solid-dosage, but not for 'high risk' parenterals. Examples necessary. Applies to the three basic 'validations' – process (including PLC-driven operations), cleaning, analytical.   [20 marks]

10. Answers: needs labelled graph of data in OCC in sigmoid-format; (a) P = 65 per cent; 2.5 per cent, 22.5 per cent approximately (± 2 per cent in value) depending on accuracy of fit [7 marks]. (b) Sampling plan – batch representation. Predictive testing with rationale [3 marks]. (c) Master plan = URS, DQ to PQ, action: role of QP, top-down or bottom-up validation strategies (conclude with VMR). ISO compliance and worst case testing [10 marks].

11. Answers: (a) Draw QC chart, label points of concern (high value). General process is . . .in order . . . ? Second improvement seems to work. Process lies outside 3-sigma at batch 25, and possibly another between 20 and 30. This lies outside acceptable limits. Process seems to inherently possess significant statistical variation if the sporadic changes in direction cannot be explained by systematic error from operators. Notions of process bias? Equipment change or revisiting. Point of change to process – QP would be required to validate and make appropriate records. Testing follows – process/product [15 marks]. (b) At the point of a minor change to process the QP would be required to circulate a change control document. If the change is not acceptable (physical chemical and conformity tests) batches to be scrapped, possibility of reworking   [5 marks].

References

Advanstar Communications (2003) *Pharmaceutical Technology Europe*, **May**, 42.

Advanstar Communications (2005) Internet at www.ptemag.com, accessed 2006.

Ahuja, S. and Scypinski, S. (2001) *Handbook of Modern Pharmaceutical Analysis*, **3**, Academic Press, New York.

Allen, T.M. (2002) *Nature Reviews Cancer*, **2** (10), 750.

Amersham plc (2006) Internet at www.amersham.com, accessed 2006.

Amir-Aslani, A. and Negassi, S. (2006) *Technovation*, **26**, 573–582.

Anik, S.T. (2002) *Pharmaceutical Technology Europe*, **October**, 12–14.

Ansel, H.C., Allen, L.V. and Popovich, N.G. (1999) *Pharmaceutical Dosage Form and Drug Delivery*, Lippincott, Williams & Wilkins, Philadelphia, 244–262; 346–396.

Balbach, S. and Korn, C. (2004) *International Journal of Pharmaceutics*, **275**, 1–12.

Banker, G.S. and Rhodes, C.T. (1979) *Modern Pharmaceutics*, Marcel Dekker, New York, 263–357.

Becher, P. (2001) *Emulsions: Theory and Practice*, Oxford University Press, UK.

Benoliel, M.J. (1999) *Trends in Analytical Chemistry*, **18**, 632–638.

Billany, M. (2002) in *Pharmaceutics: the Science of Dosage Form Design* (Aulton, M.E., ed.), Churchill-Livingstone, London, 334–359.

Black, S.A. and Porter, L.J. (1996) *Decision Science Journal*, **27** (1), 1–22.

Bloomfield, S.F. and Baird, R.M. (eds.) (1996) *Microbial Quality Assurance in Pharmaceuticals, Cosmetics and Toiletries*, Taylor and Francis, London.

Bourget, P., Perello, L. and Demirdjian, S. (2001) *Pathologie Biologie*, **49** (1), 86–95.

Breitenbach, J. (2002) *European Journal of Pharmaceutics and Biopharmaceutics*, **54**, 107–117.

Buncher, C.R. and Tsay, J.-Y. (1994) *Statistics in the Pharmaceutical Industry*, 2nd Edition, Marcel Dekker, New York.

Burns, H.D. (1978) in *The Chemistry of Radiopharmaceuticals* (Heidel, N.D., Burns, H.D. Honda, T. and Brady, L.W., eds.), Masson Publishing (Abacus Press), USA, 35–52.

Cambridge Healthcare and Biotech (2005) Internet at www.chandb.com, accessed 2006.

Carleton, F.J. and Agalloco, J.P. (1999) *Validation of Pharmaceutical Processes*, 2nd Edition, Marcel Dekker, New York.

Carpe Diem Publishers (2004) *Pharmaceutical Formulation and Quality*, **August/September**, 14–15.

Case, F. (2006) *Chemistry World (UK)*, **July**, **3** (7) 36–40.

Chattopadhay, P.K., Price, D.A., Harper, T.F., Betts, M.R., Yu, J., Gostick, E., Perfetto, S.P., Guefert, P., Koup, R.A., De Rosa, S.C., Bruchez, M.P. and Roederer, M. (2006) *Nature Medicine*, **12** (8) 972–977.

Chzanowski, F.A. (2006) *Pharmaceutical Formulation and Quality*, **April/May**, 58–60.

Cledera-Castro, M., Santos-Montes, A.M. and Izquierdo-Hornillos, R. (2006) *LC-GC Europe*, **July**, 424–425.

Colombo, P., Bettini, R., Santi, P. and Peppas, N.A. (2000) *Pharmaceutical Science & Technology Today*, **3** (6), 198–204.

Conzen, P.F. (2005) *Best Practice & Research: Clinical Anaesthesiology*, **17** (1): 29–46.

Crawford, C.M. and Di Benedetto, C.A. (eds.) (2006) *New Products Management*, 8th Edition, McGraw-Hill, Boston.

Crespi, R.S. (1999) in *Biopharmaceuticals, an Industrial Perspective* (Walsh, G. and Murphy, B., eds.), Kluwer Academic Publishers, Netherlands, 249–267.

Crowley, R. and FitzGerald, L.H. (2006) *Toxicology* **221**, 9–16.

Cundell, A. (2004) *Pharmaceutical Technology*, **June**, **28** (16), 56–61.

DEA (2006) Internet at www.usdoj.gov/dea, accessed 2008.

Dejaegher, B. and Heyden, Y.V. (2006) *LC-GC Europe*, **July**, 418–420.

Deshpande, P.B. (1998) *Hydrocarbon Processing*, **77** (4), 55–61.

Doblhoff-Dier, O. and Bliem, R. (1999) *Trends in Biotechnology*, **17**, 266–270.

Drews, J. and Ryser, S. (1997) *Drug Discovery Today*, **2** (9), 365–372.

Duncan, R. (2005) *Nano Today*, **August**, 16–17.

Eggins, B.R. (2003) *Chemical Sensors and Biosensors*, John Wiley and Sons, Chichester, UK.

EMEA (2006) Internet at www.emea.eu.int, accessed 2006.

European Parliament (2005) Internet at www.europa.eu.int, accessed 2006.

European Parliament (2006) EU Council Directive 89/343/EEC, Article 2.

European Patent Office (2008) Internet at www.epo.org, accessed 2008.

Fahmy, T.M., Fong, P.M., Goyal, A. and Saltzman, W.M. (2005) *Nano Today*, **August**, 18–26.

FDA (1996) International Conference on Harmonisation: final guidelines on stability testing of biotechnology/biological products; availability, *Federal Register*, **61**, 36466–36469.

FDA (2001) ICH. Q7A Good manufacturing practice for active pharmaceutical ingredients, *Federal Register*, **September**, **66** (186), 49028–49029.

FDA (2003) Good manufacturing practice for finished pharmaceuticals, *Code of Federal Regulations, Part 211, Title 21*, **4** (1), US GPO.

FDA and GSK (2006) Internet at www.fda.gov, accessed 2008; Internet at www.gsk.com/about, accessed 2008.

Feldman, S.R., Sangha, N. and Setaluri, V. (2000) *Journal of the American Academy of Dermatology*, **42** (6), 1017–1020.

Florence, A.T. and Attwood, D. (1998) *Physicochemical Principles of Pharmacy*, Macmillan, London, 404–414.

Floyd, A.F. (1999) *Pharmaceutical Science and Technology Today*, **2** (4), 134–143.

Forster, A., Rades, T. and Hempenstall, J. (2002) *Pharmaceutical Technology Europe*, **October**, 27–37.

Freeman, C., Hebbes, A. and Tippins, R. (2003) *Biotechnology International*, 15(5), 26.

Friedli, D., Kappeier, W. and Zimmermann, S. (1998) *Pharmaceutica Acta Helveticae*, 72(6), 343–348.

Fujii, T. (2002) *Microelectronic Engineering*, 61–62, 907–914.

Gewin, V. (2006) *Nature*, 443 (7108), 137.

Gibson, M. (ed.) (2004) *Pharmaceutical Formulation and Preformulation*, Interpharm CRC Press, Boca Raton, 157–173.

Goodwin, J. (2004) *Colloids and Interfaces with Surfactants and Polymers*, John Wiley and Sons, Chichester, 177–237.

Graumann, K. and Premstaller, A. (2006) *Biotechnology Journal*, 1, 164–186.

Greaves, P. (2006) *Special Feature: The Vision*, Bovis Lend Lease, Trafford, UK, 12–13.

Greenwood, P.A. and Greenaway, G.M. (2002) *Trends in Analytical Chemistry*, 21 (11), 726.

Gregoriadis, G. (1973) *FEBS Letters*, 36, 292.

Guy, R.H., Hadcraft, J., Kellaway, I.W. and Taylor, M. (1982) *International Journal of Pharmaceutics*, 11, 199–208.

Halstensen, M., de Bakker, P. and Esbensen, K.H. (2006) *Chemometrics and Intelligent Laboratory Systems*, 84, 88–97.

Hanlon, G. (2002) in *Pharmaceutics – The Science of Dosage for Design* (Aulton, M., ed.), Churchill-Livingstone, London, 599–622.

Harris, D.C. (1999) *Quantitative Chemical Analysis*, 5th Edition, WH Freeman, New York, 51–68.

Harris, DC (2003) *Quantitative Chemical Analysis*, 6th Edition, WH Freeman, New York, 720–739.

Hausman, D.S., Cambron, R.T. and Sakr, A. (2005) *International Journal of Pharmaceutics*, 299, 19–33.

Hesselwood, S.R. (1990) in *Textbook of Radiopharmacy Theory and Practice* (Samson, C.B., ed.), Gordon and Breach Scientific Publishers, UK, 101–112.

Hiemenz, P.C. and Rajagopalan, R. (eds.) (1997) *Principles of Colloid and Surface Chemistry*, 3rd Edition, Marcel Dekker, New York, 145–192; 355–404.

Hodges, N. (2002) in *Pharmaceutics – The Science of Dosage for Design* (Aulton, M., ed.), Churchill-Livingstone, London, 623–642.

Holm-Nielsen, J.B., Dahl, C.K. and Esbensen, K.H. (2006) *Chemometrics and Intelligent Laboratory Systems*, 83, 114–126.

Hopley, L. and van Schalkwyk, J. (2006) Internet at www.anaesthetist.com/anaes/drugs, accessed 2006.

Hora, M.S. and Chen, B.-L. (1999) in *Biopharmaceuticals, an Industrial Perspective* (Walsh, G. and Murphy, B., eds.), Kluwer Academic Publishers, Netherlands, 217–248.

Howbrook, D.N., van der Valk, A.M., O'Shaughnessy, M.C., Sarker, D.K., Baker, S.C. and Lloyd, A.L (2003) *Drug Discovery Today*, 8 (14), 642–651.

Hoyle, D. (2006) *ISO 9000: Quality Systems Handbook*, 5th Edition, Butterworth-Heinemann, London, pp 686.

Hsu, Y.-Y., Gresser, J.D., Trantolo, D.J., Lyons, C.M., Gangadharam, P.R.J. and Wise, D.L. (1996) *Journal of Controlled Release*, 40 (3), 293–202.

Huang, X., Tanojo, H., Len, J., Deng, C.M. and Krochmal, L. (2005) *Journal of the American Academy of Dermatology*, 53 (1), S26–S38.

ICRP (1977) Principles of Handling in *International Commission for Radiological Protection*, ICRP Publications, 26.

Ishizuka, H., Waki, Y., Horizuti, M., Ishikura, C. and Awazu, S. (1995) *International Journal of Bio-Medical Computing*, **38** (2), 167–176.

ISO (2000) EU Good Manufacturing Guide, Chapter 9; ISO9001:2000.

ITSM (2005) Internet at www.itil-itsm-worrld.com/sigma.htm, accessed 2006.

Johnson, D. (2003) *Pharmaceutical Technology Europe*, **December**, 57–61.

Kellner, R., Mermet, J.-M., Otto, M. and Widmer, H.M. (eds.) (1998) *Analytical Chemistry*, Wiley-VCH, New York, 41–68; 339–358; 709–824; 857–859.

Kennedy, T. (1997) *Drug Discovery Today*, **2** (10), 436–444.

Klang, S.H., Parnas, M. and Benita, S. (1998) in *Emulsions and Nanosuspensions for the Formulation of Poorly Soluble Drugs* (Müller, R.H., Benita, S. and Böhm, B.H.L., eds.) Medpharm GmbH Scientific Publishers, Stuttgart, 31–65.

Klotz, U. and Schwab, M. (2005) *Advanced Drug Delivery Reviews*, **57**, 267–279.

Kolarik, W.J. (1995) *Creating Quality: Concepts, Systems, Strategies, and Tools*, McGraw-Hill, London.

LeBlanc, D.A. (2000) *Pharmaceutical Technology*, **24** (10), 160–168.

Loftus, B.T. and Nash, R.A. (1984) *Pharmaceutical Process Validation*, Marcel Dekker, New York.

Mann, D.L. (2006) *Nature Medicine*, **12** (8), 881.

Martin, A.N. (1993) *Physical Pharmacy*, 4th Edition, Williams & Wilkins, Baltimore, 77–100; 477–497.

MCA (1997) *Rules and Guidance for Pharmaceutical Manufacturers and Distributors 1997*, MCA, Stationery Office Ltd, 3–177.

McDowall, R.D. (1999) *Analytica Chimica Acta*, **391**, 149–158.

Melethil, S. (2006) *Life Sciences*, **78**, 2049–2053.

MHRA (2002) *Rules and Guidance for Pharmaceutical Manufacturers and Distributors 2002*, Medicines Control Agency, HMSO, Norwich, 3–177.

MHRA and Roche (2006) Internet at www.mca.gov.uk, accessed 2008; Internet at www.roche.com/home, accessed 2008.

Moghimi, H.R. (1996) *International Journal of Pharmaceutics*, **131**, 117–120.

Mollah, A.H. (2004) *BioProcess International*, **2** (9), 28–35.

Moritz, A. (2005) *BioProcess International*, **3** (2), 28–38.

Mudhar, P. (2006) *Pharmaceutical Technology Europe*, **18** (9), 20–25.

Muller, K.M., Gempeler, M.R., Scheiwe, M.-W. and Zeugin, B.T. (1996) *Pharmaceutica Acta Helveticae*, **71** (6), 421–438.

Müller, R.H. and Böhm, B.H.L. (1998) in *Emulsions and Nanosuspensions for the Formulation of Poorly Soluble Drugs* (Müller, R.H., Benita, S. and Böhm, B.H.L., eds.) Medpharm GmbH Scientific Publishers, Stuttgart, 151–160.

Munden, R., Everitt, R., Sandor, R., Carroll, J. and DeBono, R. (2002) *Pharmaceutical Technology Europe*, **October**, 66–72.

Nakao, R., Furutuka, K., Yamaguchi, M. and Suzuki, K. (2006) *Nuclear Medicine and Biology*, **33** (3), 441–447.

Närhi, M. and Nördstrom, K. (2005) *European Journal of Pharmaceutics and Biopharmaceutics*, **59**, 397–405.

Naylor, S. (2004) *Drug Discovery Today: Targets*, **3** (2, suppl. 1), 1–2.

Nazzal, S. and Khan, M.A. (2006) *International Journal of Pharmaceutics*, **315**, 110–121.

Neville, J. (2005) *LabPlus International* **February/March**, 27–29.

Nielsen, S.H. and Gohla, J. (1998) in *Emulsions and Nanosuspensions for the Formulation of Poorly Soluble Drugs* (Müller, R.H., Benita, S. and Böhm, B.H.L., eds.) Medpharm GmbH Scientific Publishers, Stuttgart, 67–78.

Norris, P. and Baker, M. (2003) *Pharmaceutical Technology Europe*, **November**, 30.

Oakland, J.S. (2000) *Total Quality Management: Text with Cases*, Butterworth-Heinemann, UK.

Oropesa, P., Serra, R., Gutierrrez, S. and Hernandez, A.T. (2002) *Applied Radiation and Isotopes*, **56** (6), 787–795.

Ottosson, S. (2004) *Technovation*, **24** (3), 207–217.

Owunwanne, S.R.A., Patel, M. and Sadek, S. (1995) *The Handbook of Radiopharmaceuticals*, Chapman and Hall Medical, London, 1–99.

Parker, B.M. (2006) *Cleveland Clinic Journal of Medicine*, **73** (suppl. 1), S13–S17.

Poe, T.A. (2003) *Drug Discovery Today*, **8** (13), 570–573.

Powell-Evans, K. (2002) *Pharmaceutical Technology Europe*, **September**, 60–65.

Pritchard, E. (ed.) (1995) *Quality in the Analytical Chemistry Laboratory*, Analytical Chemistry by open Learning, UK, 1–66.

Pyzdek, T. (2001) *The Complete Guide to Six Sigma*, McGraw-Hill, USA.

Rahman, S. and Bullock, P. (2005) *Omega*, **33**, 75–83.

Rang, H.P. (ed.) (2006) *Drug Discovery and Development Technology in Transition*, Churchill-Livingstone, London.

Reich, G. (2005) *Advanced Drug Delivery Reviews*, **57**, 1109–1143.

Reinertsen, D.G. (1997) *Managing the Design Factory, A Product Developer's Toolkit*, The Free Press, London, pp 540.

Riess, J.G. and Krafft, M.P. (1998) *Biochimie*, **80**, 489–514.

Robertson, G.L. (1993) *Food Packaging, Principles and Practice*, Marcel-Dekker, New York.

RPSGB (2008) Internet at www.pharmj.com, accessed 2008.

Salvage, F. (2002) *Chemistry in Britain*, **May**, 30–34.

Samson, D. and Terziovski, M. (1999) *Journal of Operations Management*, **17**, 393–409.

Sarker, D.K. (2004) *Business Briefing: Future Drug Discovery*, 38–41; www.touch briefings.com/pdf/890/PT04_sarker.pdf, accessed 2006.

Sarker, D.K. (2005a) *Current Drug Delivery*, **2** (4), 297–310.

Sarker, D.K. (2005b) *Current Nanoscience*, **1** (2), 157–168.

Sarker, D.K. (2006) *Mini Reviews in Medicinal Chemistry*, **6** (7), 793–804.

Schacter, B. (2006) *The New Medicines: How Drugs are Created, Approved, Marketed and Sold*, Praeger Publishers, Connecticut, 114–270.

Schuster, M., Nechansky, A., Loibner, H. and Kircheis, R. (2006) *Biotechnology Journal*, **1**, 138–147.

Selkirk, A.B. (1998) *Pharmaceutical Science & Technology Today*, **1** (1), 8–11.

Shappiro, J. and Maddin, S. (1996) *Clinics in Dermatology*, **14** (1), 123–128.

Sharp, J. (2000) *Quality in the Manufacture of Medicines and Other Healthcare Products*, Pharmaceutical Press, London.

Sharp, J. (2002) *Industrial Pharmacist*, **26** (**September**), 4–6.

Shaw, D.J. (1992) *Introduction to Colloid and Surface Chemistry*, Butterworth-Heinemann, Oxford, 174–209; 262–276.

Sinko, P.J. (2006) *Martin's Physical Pharmacy and Pharmaceutical Sciences*, 5th Edition, Lippincott Williams & Wilkins, Philadelphia, 499–530; 585–680.

Six Sigma Tutorial (2005) Internet at www.sixsigmatutorial.com/Six-Sigma/Six-Sigma-Tutorial.aspx, accessed 2006.

Skoog, D.A., Holler, F.J. and Nieman, T.A. (1998) *Principles of Instrumental Analysis*, Harcourt Brace Publishers, Philadelphia.

Skoog, D.A., West, D.M., Holler, F.J. and Crouch, S.R. (2000) *Analytical Chemistry: an Introduction*, 7th Edition, Harcourt, USA, 471–492.

Slater, S. (1999) in *Biopharmaceuticals, an Industrial Perspective* (Walsh, G. and Murphy, B., eds.), Kluwer Academic Publishers, Netherlands, 311–335.

Smith, A. (1999) *Pharmaceutical Science and Technology Today*, **2** (6) 225–227.

Snee, R.D. (1986) *Quality Progress*, **August**, 25–31.

Snee, R.D. (1990) *The American Statistician*, **44**, 116–120.

Snelling, C.F.T. (1980) *Burns*, **7** (2), 143–149.

Sonnergaard, J.M. (2006) *European Journal of Pharmaceutics and Biopharmaceutics*, **63**, 270–277.

Speiser, P.P. (1998) in *Emulsions and Nanosuspensions for the Formulation of Poorly Soluble Drugs* (Müller, R.H., Benita, S. and Böhm, B.H.L., eds.) Medpharm GmbH Scientific Publishers, Stuttgart, 15–28.

Tambuyzer, E. (2002) *Biopharmaceuticals Europe*, **September**, 19–22.

Taverners, I., De Loose, M. and Van Bockstaele, E. (2004) *Trends in Analytical Chemistry*, **23** (8), 535–552.

Thomas, A.C. and Campbell, J.H. (2004) *Journal of Controlled Release*, **100**, 357.

Tian, J., Liu, J., Hu, Z. and Chen, X. (2005) *Bioorganic and Medicinal Chemistry*, **13**, 4124.

Torchilin, V.P. (2001) *Biochimica et Biophysica Acta*, **1511** (2), 397.

UK HSE (1985) UK Ionising Radiation Regulations 1985.

UKRG (2006) Internet at www.ukrg.org.uk/rphandbook.diluent2.htm (top of page 57), accessed 2006.

Underwood, E (1995) *International Biodeterioration & Biodegradation*, **36** (3-4), 449–457.

University of Florida (2006) Internet at www.cop.ufl.edu/safezone/prokai/pha5100/Emulsion.htm, accessed 2006.

Valtcheva-Sarker, R.V., O'Reilly, J.D. and Sarker, D.K. (2007) *Recent Patents on Drug Delivery & Formulation*, **1** (2), 147–159.

Verghese, G. (2003) *Pharmaceutical Technology Europe*, **November**, 22–29.

Vickers, M.D., Morgan, M. and Spencer P.S.J. (1991) *Drugs in Anaesthetic Practice*, 7th Edition, Butterworth-Heinemann, Oxford, 56–217.

Vogleer, J. and Boekx, J. (2003) *Pharmaceutical Technology Europe*, **May**, 25–30.

Wagner, R. (2006) *Pharmaceutical Formulation and Quality*, **December/January**, 62–66.

Walsh, G. and Murphy, B. (eds.) (1999) *Biopharmaceuticals, an Industrial Perspective*, Kluwer Academic Publishers, Netherlands.

Wang, N., Wu, Q., Xiao, Y.M., Chen, C.X. and Lin, X.F. (2005) *Bioorganic and Medicinal Chemistry*, **13**, 2667.

Washington, C. (1998) in *Emulsions and Nanosuspensions for the Formulation of Poorly Soluble Drugs* (Müller, R.H., Benita, S. and Böhm, B.H.L., eds.) Medpharm GmbH Scientific Publishers, Stuttgart, 101–117.

Washington, C. (1990) *International Journal of Pharmaceutics*, **58**, 1–12.

Webster, G. K., Kott, L. and Maloney, T.D. (2005) *Journal of the Association for Laboratory Automation*, **10** (3), 182–193.

Webster, J. (ed.) (1995) *Medical Instrumentation, Application and Design*, 2nd Edition, John Wiley and Sons, Chichester.

Weckwerth, W. and Morgenthal, K. (2005) *Drug Discovery Today*, **10** (22), 1551–1558.

Whalen, F.X., Bacon, D.R. and Smith, H.M. (2005) *Best Practice & Research: Clinical Anaesthesiology*, **19** (3), 323–330.

Wiley-VCH and FDA (2006) *Biotechnology Journal*, **1**, 124–126; Internet at www.fda.gov/cber/efoi/approve.htm, accessed 2006.

Yonzon, C.R., Stuart, D.A., Zang, X., McFarland, A.D., Haynes, C.L. and Van Duyne, A.D. (2005) *Talanta*, **67**, 438–448.

Index